The Essential Guide to the Dubai Real Estate Market

This is the first book to fully present, analyse and interpret the Dubai real estate market. Dubai is fast becoming one of the world's most attractive places to invest in real estate and this book examines the market from three interlinked sectors that drive its performance: occupiers, investors and developers. It examines the market's historical growth and lays the foundations to examine future trends. The book provides a synopsis of Dubai's market practices, economic trends and social change that impacts the value of real estate. Chapters also debate issues such as property investment, house price performance, local valuation practices, spatial planning, the economics of the city, market practices and regulation, property-led economic growth and future trends such as sustainability and digitalization. This book offers a comprehensive, in-depth and up-to-date account of the Dubai property market and presents a full assessment of the investment potential of Dubai real estate. It is a must read for students, academics and real estate professionals interested in this fascinating real estate market that has implications for both Dubai and wider GCC markets as well as the international investment market and senior professionals who come to work in the region.

Michael Waters is Associate Professor in Real Estate at Heriot-Watt University in Dubai. He has been working in real estate academia for over 20 years and is a member of the Royal Institution of Chartered Surveyors. He is a prominent expert in real estate in Dubai and makes regular contributions to media publications.

International Real Estate Markets

The Chinese Real Estate Market
Development, Regulation and Investment
Junjian Cao

Real Estate in Italy
Markets, Investment Vehicles and Performance
Guido Abate and Guiditta Losa

Real Estate Markets and Development in South America
Understanding Local Regulations and Investment Methods in a Highly
Urbanised Continent
Claudia Murray, Eliane Monetti and Camilla Ween

Real Estate in South Asia
Edited by Prashant Das, Ramya Aroul and Julia Freybote

Real Estate in Central America, Mexico and the Caribbean
Claudia Murray, Camilla Ween, Yadira Torres and Yazmin Ramirez

The Real Estate Market in Ghana
An Emerging Market in Sub-Saharan Africa
Wilfred K. Anim-Odame

Understanding African Real Estate Markets
*Edited by Aly Karam, François Viruly, Catherine Kariuki and Victor
Akujuru*

The Essential Guide to the Dubai Real Estate Market
Michael Waters

The Essential Guide to the Dubai Real Estate Market

Michael Waters

Routledge
Taylor & Francis Group

LONDON AND NEW YORK

Designed cover image: © Getty Images

First published 2023
by Routledge
4 Park Square, Milton Park, Abingdon, Oxon OX14 4RN

and by Routledge
605 Third Avenue, New York, NY 10158

Routledge is an imprint of the Taylor & Francis Group, an informa business

British Library Cataloguing-in-Publication Data
A catalogue record for this book is available from the British Library

ISBN: 978-1-032-03357-0 (hbk)
ISBN: 978-1-032-03356-3 (pbk)
ISBN: 978-1-003-18690-8 (ebk)

DOI: 10.1201/9781003186908

Typeset in Times New Roman
by codeMantra

To Dee, Lily & Phoenix

Contents

Preface

The real estate market in Dubai is relatively young when compared to other international cities, having opened to foreign real estate investment only in the last 20 years (in 2002). The market as with many globally is multi-dimensional and complex. Yet, Dubai is fast becoming one of the world's most attractive places to invest in real estate. There are numerous reasons for this growth and success. In recent years, Dubai has developed a real estate market with stronger legislation and now has a rich amount of data from which comprehensive analysis can be done. The first edition of this book provides the foundations to understanding many of the key parts of the market and industry practices. It will help students and general readers piece together the important aspects essential to better understand Dubai's real estate.

An academic book about Dubai real estate is long overdue, yet could take many forms. I have chosen to examine the market from a wide perspective. I have also chosen to make comparisons with other international markets so that readers who are perhaps familiar with other real estate jurisdictions can put Dubai into context. The writing of this book has also taken a frank approach to some of its analysis to highlight where the market still has room to mature and develop. I have been fortunate enough to have spent an extensive part of my career working in Dubai to make these comparisons. I also do it with an impartiality as a real estate academic. I am also a qualified Chartered Property Surveyor (MRICS) and have always been keen to relate theories to real-world practices. This book is therefore a combined result of both my academic and professional observations. The aims of this book are discussed in more detail below.

The purpose of this book *The Essential Guide to the Dubai Real Estate Market* is to provide background reading for students of real estate, business, finance and land economics to whom an insightful understanding of Dubai as both a city and a new global case study would be of great interest. The lack of such a book was noted when I arrived in Dubai in 2009 and as I continued to teach real estate courses in Dubai and speak with students, young practitioners and real estate agents it appeared that a concise reference and academic insight into Dubai would be beneficial.

Over the last 20 years, teaching across a wide range of the real estate programme syllabus has equipped me with vast information and comparative analysis on Dubai. It has also been a very exciting time to have been based in Dubai to observe the city grow and develop and be on par with some of the world's top cities. Its global competitiveness and pace of growth has been remarkable to observe. The city has provided me with a wonderful array of experiences and this book in one way is trying to reflect those observations in a general yet concise guide. I hope this book will be read by a wide audience. It has not been written to complicate or be critical of how things should or should not be done, but I have tried to explain things from an informed perspective using real-life data and real-world examples from a wide academic base; via a number of industry and practitioner observations and perhaps even a synopsis of the lively classroom discussions I have had with my students. Therefore I hope that this book will also be useful to visiting practitioners, new entrants to Dubai as well as a wider public who might simply have a curiosity of Dubai and its real estate market.

The aims and objectives of this book

This is the first book to fully present, analyse and interpret the Dubai real estate market. It begins by examining the market's historical growth and lay foundations to examine future trends. The book provides a synopsis within one publication that benefits global real estate students and practitioners on Dubai's market practices; economic trends and changes that have and will continue to impact the value of real estate. This book also debates issues such as property investment; house price performance; local valuation practices; real estate development; the economics of the city; market practices and regulations; and future trends such as sustainability and the development of Dubai as a smart city. The book offers a comprehensive, in-depth and up-to-date account of the Dubai property market and presents a full assessment of the potential Dubai has to becoming a top-tier international real estate market. External to the student population studying real estate in Dubai, there would also be interest for new entrants graduates and senior professionals who come to work in Dubai's real estate market who are seeking a comprehensive overview of a range of real estate practices and regulations ranging from valuation to leasing and letting; purchase and sale and corporate real estate. The book can also serve as a comprehensive guide for other global students and professionals working in Dubai who can use the material as a suitable study guide to local market practices. I also hope that it is an easy-pick up and read for Dubai residents or potential home buyers who are keen to understand the market performance; appraisal techniques and buying and selling processes relevant to the more practical and applied chapters I have written for such purpose (Chapter 2,3,4,8 and 9 in particular).

This book represents the first academic publication of its kind and in some part is a reflective piece based from my perspective having been in Dubai since 2009. As the book examines topics it tries to present comparative analysis and a range of insights from leading stakeholders who share their experiences of working in Dubai's property market. This makes for an interesting read and I feel allows students and practitioners a good point of reference on a wide range of topics that are important for those who are seeking a new global case study to evaluate; or to the new graduates who come from other global markets who have ambitions of working in Dubai and need a comprehensive guide to how the market is structured and how it operates. Also to the wide range of other professional agents who work in real estate sales and leasing who seek an overview of how the city has developed and discussions based around data; market commentary and key trends that are shaping the new opportunities that exist in real estate investment – both for individual and institutional investors. Dubai leads the Middle East region in many of its accolades from transactional activity; to data transparency and seeking new ways to transact through blockchain technology and digital currencies; 3D printing of buildings; and a continuation of its truly ambitious plans to be the world's most futuristic city. I have tried to do the city justice by writing this book in an engaging manner that capitalizes on just how exciting Dubai is as a place to visit, work and live. The purpose of this book is to ensure that students and practitioners have a ready guide to a range of topical aspects ranging from property valuation to leasing and letting; real estate development and asset management, without intending that this book will teach the pre-existing academic theories, but to showcase the similarities and differences in market practices. There is a myriad of suitable textbooks that detail the theoretical underpinnings of these topics, but what this book does offer is a new global case study in which to frame the development of these in Dubai.

No single book can be a complete authority on a subject and this book is no exception to this. Like us all I have my professional competencies and also I have my limitations. Therefore whilst I reference certain areas of legislation or legal systems I am not a lawyer. My interest lies in setting these frameworks out and analyzing them in the context of how they relate to a range of professional areas including property management and valuation. Again these analyses are more comparative than pioneering theoretical discussions. I would expect readers of this book to gain a better understanding to Dubai's real estate market and its operations.

To end, it must be clearly expressed here that this book is not written as professional advice or investment advice; they simply represent my thoughts. Anyone using the contents of this book to make financial decisions is doing so entirely at their own risk. Neither the author nor any other party connected with this publication can accept responsibility for the results of any actions taken as a result of reading it.

The structure of the book

The chapters within this book have been carefully planned to provide a range of topical accounts of Dubai's real estate growth and development. Chapter 1 starts the analysis by examining the historical growth of Dubai, answering the question "How did Dubai become a leading global real estate market?". It is a market composed of numerous polycentric districts across one city. It provides an overview of the economic and geographic forces that have helped shape Dubai and its real estate market, covering the growth and development over the last 50 years. It provides context to Chapters 2–5, by discussing the evolution and maturity over the "freehold" period, from 2002, when Dubai first began to allow international investors to buy and invest in Dubai. A summary of market-influencing legislation and data to enhance market efficiency is provided. Chapter 2 is an overview of this property market activity over the last 20 years, setting both a theoretical context and a range of market commentary on the growth and development of real estate investment in Dubai. Chapters 3 and 4 then take these discussions and examine the residential sector in more detail, taking a more general approach to investment theory, then going onto discuss the internationalization of capital into Dubai both from an individual to institutional level. Chapter 4 takes on the role of providing a suitable analysis for individual investors to buy into Dubai's residential market, offering insights and commentary to how and why investing in Dubai's real estate market can work well for an international investor, examining the benefits of diversification to the currency returns over long-term investment periods. A range of practicalities such as the different forms of finance and the impact of leverage in property cycles is brought out in these discussions. Chapter 5 moves away from the residential sector and provides a similar analysis for the commercial sector. Chapter 6 is an overview of sales and leasing practices in Dubai that provides the context to occupying and investing in Dubai. Chapter 7 puts leasing practices into a real-world context and evaluates the decisions made by corporate occupiers in Dubai. Aside from these commercial and legal interests, property data are fundamentally important for a wide range of real estate professionals and Chapter 8 provides an understanding of the forms of property data available in Dubai. From this comes an explanation of property valuations themselves. Chapter 9 examines the challenges faced in property valuations and highlights the nuances of valuing real estate in Dubai versus international markets. The process of property development is important to understand (Chapter 10) and new development patterns point to how the city will evolve in the future. Developers in Dubai have been historically faced with a range of development risks and challenges which are brought out in the discussions. A key challenge for any major global city looking forward is its ability to engage with the range of climatic risks that are facing the sector, and Chapter 11 highlights the progress of sustainability in real estate in Dubai. It demonstrates what is currently developed and what is being developed to comply with a significant global rise in ESG and environmentally responsible real estate. The final chapter (Chapter 12) is a future-forward examination of how Dubai is set to become one of the world's smartest cities. How will Dubai look in the next 20 years?

Acknowledgements

Despite this book being published by a single author, there are many people to thank who have made important contributions to this book.

I would like to thank the Routledge team for their patience with me in getting this 1st edition into print. All their expertise and guidance made the book-writing process a lot more straightforward.

I would like to thank the main contributors, Mike Wing (who co-wrote Chapter 10) and Tim Shelton (who wrote Chapter 12) who contributed a significant amount of expertise and time into those respective chapters. Further industry insights were provided by Taimur Khan in Chapter 2 (Box 2.2); Andrew Baum, for his shared case study data (*from Real Estate Investment, A Strategic Approach: fourth edition, Routledge*) which has been reworked into a new context for my own analysis in Chapter 3 (Box 3.1) and Faisal Durrani in Chapter 5 (Box 5.1). I would also like to thank Simon Townsend (who provided feedback to Chapter 5); Andrew Love and Erik Volkers (who supplied, where noted, industry data in Chapters 5 and 10); Zhann Jochinke for sharing relevant data from Property Monitor; Vidhi Shah (who shared her insights to the Case Examples 1–3 in Chapter 9); and Angela Abeidat for sharing and discussing a range of useful information regarding the Expo 2020 site.

The knowledge shared by those above and their contributions of supporting thoughts and information have enriched the quality of my written work in key parts of this book.

Part A

Contextualizing the Dubai real estate market

1 How did Dubai become a leading global real estate market?

This book has not been written to be historical, yet it is important to highlight the economic roadmap that Dubai has taken to get to where it is today. For those interested in real estate and the evolution of cities, Dubai is a somewhat unique case study, yet perhaps its origins of development are no different to other major global cities. Dubai, like many others, has been built off commerce. It evidently started out in very similar ways to other major international cities, it had something to offer the world. Dubai has developed remarkably quickly and remarkably well by most metrics. While other commentators go far in back its evolution, the purpose of the first chapter of the book is to draw on the main economic forces that created Dubai. Most suitably, this analysis focuses on a reflection of the last 50 years of independence, with the UAE celebrating its 50th national day in 2021. The final chapter looks forward to the next 50 years (Chapter 12). After studying this chapter, you will be able to:

- Understand the macroeconomic development of Dubai
- Understand the sequential urban development of Dubai
- Provide an outline of the main economic forces that have helped shape Dubai
- Evaluate Dubai's development against other global real estate markets

This chapter begins by providing an overview of the historic economic development of Dubai.

Brief historical development: The economic forces that shaped Dubai

Dubai's origins began from a strategic geographical trade position. A key economic force for the Emirate and its historical development was its geographical advantage as a trading point between East and West. Dubai initially grew as a port city. During its "Trucial State" (1820–1971), its proximity between India and Europe was advantageous, particularly in relation to the trade of gold. Beyond geography, the governance of Dubai made

DOI: 10.1201/9781003186908-2

some strategic economic decisions that solidified Dubai's position as a trade post. For instance, during the 1890s a series of tax concessions were granted to foreign traders, which meant Dubai soon became the base of operations for regional trade. The dredging of Dubai Creek provided access for larger vessels and the city soon became known for import and re-export of goods. Dubai's economic development grew from these foundations. Within the economic development that has helped shape Dubai, we can refer to a range of forces including agglomeration; clusters and urban competitiveness. Trade and commerce are still prevalent in Dubai's economic growth. This allowed for the development in the 2000s of a large service and commercial sector (financial; real estate). Similar economic growth patterns were seen in the likes of Hong Kong and Singapore, unsurprisingly given the similar policy of low or no taxes in these jurisdictions; presence as regional financial hubs as well as geographical trade advantages. Today, Dubai is consistently ranked high in global cities. JLL's City Momentum Index placed it as the world's third-most dynamic city, taking note of its strategic global location, calling it "the crossroads of Europe, Central Asia, South Asia and Africa... with more than two-thirds of the world's population living within an 8-hour flight time" (JLL, 2020). Dubai is a truly globally connected city.

Table 1.1 presents an overview of key economic and development milestone events that have helped shape Dubai. From an examination of key events in Table 1.1, there was an economic advantage to use Dubai as a trading hub, first for commerce and more recently since the expansion of Emirates Airlines in mid-1990s, tourism and service economies. Modern Dubai has also developed a significant emphasis placed upon property-led policies via real estate development and construction.

As discussed in the earlier analysis, the key to the early development of Dubai was the port-based infrastructure that facilitated it position in global trade, during the late 1950s and throughout the 1960s and 1970s. The first spatial plan (masterplan) of Dubai was presented in 1960 by British architect, John Harris. In this plan, Dubai Creek was highlighted as Dubai's first urban centre, playing a key role as a centre of global trade and economic activity. Beyond this centre, a grid of road infrastructure and a zoning system (residential; commercial and industrial areas) were used to plot the development further out into the desert.

Since the initial urban development plan, a number of periodic revisions have been made to accommodate some of the major dynamic changes that impacted Dubai's development pathway. Notable changes in the 1971 plan included the expansion of new development along coastal areas (primarily Jumeirah). Major road transport infrastructure (namely the Sheikh Zayed Road) supported this future growth aspiration. Maktoum Bridge (1963) and Shindagha Tunnel (1967) were both built over the Creek to link the historic districts of Deira and Bur Dubai. Port Rashid, completed in 1972, and later was joined by the opening of Jebel Ali Port in the 1980s, a $2.5bn investment located towards the south of the city. This led to new development

Table 1.1 Key historical timeline of Dubai

1950	Oil discovered in the Trucial States
1960	Dubai International Airport opens. Jumeirah/Al Quoz districts established
1961	Dubai Creek dredged to support a larger global logistics trade
1963	Maktoum Bridge was built. First bridge over the creek
1966	Commercial quantities of oil discovered off coast of UAE. Dubai's first hotel was built. Dubai as a wealthy trading hub, also discovered oil. The rulers of Dubai ensured the use of oil revenues were used to support social and economic development
1967	Shindagha Tunnel becomes the second bridge built over the creek
1971	UAE becomes independent, and joins the Arab League
1972	Port Rashid opened. A major player in Dubai's international development. Subsequently in 2008 the port closed and has been outlined for residential and commercial redevelopment
1973	UAE launches its currency (AED – UAE Dirham)
1979	Sheikh Rashid Tower (now known as Dubai World Trade Centre) opened heralding the start of Dubai's rise as a top Middle Eastern trade and financial hub
1981	UAE joins the Gulf Cooperation Council (GCC)
1983	Jebel Ali Port established
1985	Emirates Airlines started operations, marking the beginning of Dubai as a top global tourism destination
1995	Dubai Investment Park opens/Jumeirah Beach Hotel
1999	Burj Al Arab, the world's tallest hotel, opens
2000	Dubai International Airport T1; Implementation of online Dubai Land Department (DLD) systems; Emirates Towers; Dubai Internet City (free zone)
2002	Foreign ownership laws ("free zones") permit the foreign ownership of real estate in Dubai
2003	Dubai formally recognized as a financial hub (World Bank/IMF)
2006	Freehold decree issued; DIFC established
2007	RERA Escrow laws established. Palm Jumeirah opens/Atlantis hotel opens (2008); Mall of Emirates opened
2008	Landlord and tenant law enacted in Dubai/Ownership of Jointly Owned Property introduced. Jumeirah Beach Residences developed; Dubai Mall opens
2009	Dubai Metro opened and became operational
2010	Burj Khalifa becomes the tallest tower in the world; Emaar Square open; Al Maktoum International Airport opens
2011	Jumeirah Lake Towers developed
2013	Dubai wins the hosting city of World Expo 2020
2014	RERA Rental index established/Mortgage caps introduced to reduce investor speculation
2015	Property transfer fees increased from 2% to 4%
2016	Dubai Creek Harbour masterplan launched
2019	Mina Rashid redevelopment announced
2020	Mortgage LTV reduced to 20% (from 25%) for expats; ICD Brookfield Place (the largest LEED office building built); Dubai South Enterprise Zone formed
2021	Dubai hosts the Expo 2020; Dubai Ain (Dubai Eye), the world's largest observation wheel opens at Bluewaters; Royal Atlantis opens

Source: Knight Frank (2021); Explorer Publishing (2009).

and expansion towards the borders of Dubai and Abu Dhabi. These developments alongside Port Rashid formed key parts of Dubai's second masterplan (Ramos, 2010). It also led to the shift of new development away from the Creek and a more south-westerly urban growth. This created a new district where large industrial and residential communities then began to develop. The historic centre of Deira and the Creek were modernised overtime yet formed one of Dubai's most well-known dense urban centres, a key focal point for Dubai, even today. The success of this huge investment in port infrastructure was also met with attractive business concessions for investors and businesses alike. With the establishment and completed development of Dubai's World Trade Centre in 1979, Dubai was seen to be open to global business, attracting new forms of international businesses, and Sheikh Zayed Road became a new corridor for commercial growth (Alshafieei, 1997). These developments marked a second major influx of foreign expat workers to Dubai (largely from Southern Asian and other Arab states), meeting the needs of the construction sector via these major infrastructure projects (Pacione, 2005).

Under the continued development vision during the 1990s, a service economy needed to be established to support the economic growth of Dubai. The "new-economy" sectors of knowledge/technology-based businesses, tourism and real estate were created (Bloch, 2010). The growth of a service sector economy in Dubai was supported with the governance of "freezones", including the likes of Dubai Internet City, established in 2000, with some of the world's best known corporate tenants including Dell, Cisco, Microsoft and Siemens, Dubai Media City (CNN, Reuters, Dow Jones) and Knowledge Village, supporting a vision towards innovation and knowledge transfer. At present, there are a total of 30 free zones in Dubai and their scope has expanded to cover new sectors such as Healthcare, Media, IT and Education. The financial services industry was boosted further with the opening of the Dubai International Financial Centre (DIFC) in 2004. These zones offered a range of financial incentives to international companies including 100% exemption from personal income tax for 50 years; 100% exemption from corporate taxes for 50 years; and 100% repatriation of profits (Jones, 2013). The UAE is the only country in the Middle East where a large number of such zones are available. As a result many international businesses chose Dubai as a base from which to access the wider Middle Eastern markets. As these new business centres developed, Dubai's population steadily grew, facilitating the development of residential and mixed-use amenities built in the vicinity.

The growth and development of "Brand Dubai" spurred tourism growth alongside the supporting infrastructure of Emirates Airlines. In 1985, there were 40 hotels in Dubai. By 2019, there were 545 hotels. The development of new hotels and tourist amenities were developed along the coastline of Dubai. The scale of real estate development had to be befitting of the scale of the ambitious plans of Dubai and it saw the development of the Palm Jumeirah, World Islands and Burj Khalifa during this time period. With increasing tourism numbers came the need to promote Dubai as a retail destination. The city's first large-scale malls, Al Ghurair City and Deira City Centre were developed in the mid-1990s. Other high-profile malls like the Mall of the Emirates and Dubai Mall opened up in 2007 and 2008. Since then the number of malls in Dubai has reached 65.

The urban development of Dubai as a trade and logistics hub is clearly apparent within its first 25–30 years of economic development. To add, Dubai's early economic and urban development was built off:

1 Advanced free zone and freeport policy (no/low tax tariffs)
2 Oil revenue-led capital infrastructure and development projects
3 Strong urban "privatized" governance and growth strategy

The third key ingredient to a successful city is governance. Dubai certainly benefits from good governance when assessments are made on its global competitiveness. Box 1.1 expands on this to highlight how governance as well as economic advantage has helped shape Dubai (see Box 1.1).

Box 1.1 Good governance: How Dubai sold itself to the World?

Key lessons from Dubai on how it has set itself up to become a leading global city include:

- Dubai is business friendly and investment ready: A long-term relationship to serve global businesses. The attractive tax concessions and foreign ownership rights as well as a reliable supply of real estate investment opportunities for overseas capital, matched by an efficient transaction process.
- Dubai as a city is run like a multi-national business: It operates like the private sector and has a branding and marketing strategy to match.
- Governance and leadership are exemplary in Dubai, benefitting from a continuous rule rather than the waves of political party motivations seen in many other global cities.
- Dubai has welcomed diverse, talented and international populations, which is a major factor when businesses decide where to locate. Open cities, like Dubai, are attractive to large multinational firms seeking access to the wider Middle East economy.
- Dubai as part of the UAE has a dollar-pegged economy and although that provides some macroeconomic policy challenges, a stable globally attracted currency like USD has offered the global investment community an attractive range of opportunities to invest and secure attractive yields, at times in the double-digits.

Economic growth in Dubai (modern era)

Since 1975, Dubai's economy has expanded by 11 times in real terms (Elhadawi and Soto, 2012). This growth was markedly observed during the 15 years between 1991 and 2008, with annual growth rates averaging 9% per annum (with its highest concentration of growth occurring post 2000). During this time the governance of Dubai's real estate sector was shifting, permitting foreign ownership in its "freehold" areas. Thompson (2016) referred to it as "state-capitalism". Dubai's regional-hub role is indisputable and has been developed over a long period through infrastructural investment (Bloch, 2010). Box 1.2 provides a synopsis of the key property development environment pre-2008, somewhat of a state-led expansion.

Box 1.2 Real estate development: The pre-2008 building boom

In the early 2000s (particularly 2004–2007), the real estate development market was being driven by excess liquidity off a sustained period of energy prices. Government-backed developers, like Emaar and Dubai Holdings, were able to leverage off high oil prices to fund new mega projects. The sheer scale of new development raised concerns over the longevity of further growth in the market at that time. Real estate consultants were calling for a phasing of this new supply to "smooth" out any supply-side shocks (Colliers, 2007), but developers had plans to capitalize on the excessive returns, which in turn attracted many new entrants to build more under the assumption that there would be enough profits for everyone. Key drivers to new real estate development included:

- There was no rental increase law present at the time so investors were freely able to raise rents without restriction (in early 2007 a 7% rental cap was temporarily introduced to curtail aggressive rental increases). Lease terms favoured the landlord disproportionately, with very few laws protecting the rights of tenants, creating a large pool of ready investors who lined up to buy what developers were building (also on the back of excess returns).
- Landlords had the upper hand and high rental returns prompted new development of both high-end residential and commercial properties. Between 2006 to 2009,there was more than a threefold increase in Dubai office space (to 5.5 million sq.m.), with Business Bay (1.6 million sq m) and Tecom (1.1 million sq m) the main source areas of new office development during this period (Colliers, 2007).
- Retail development was also at centre stage during this period as Dubai saw it necessary to diversify its economy and move away

from its dependence on oil-based revenues. Tourism and leisure spear-headed the development of new super-regional malls
like Mall of Emirates (223,000 sq m GLA), Dubai Festival City
(250,000 sq m GLA) and Dubai Mall (344,000 sq m GLA). These
new mega mall concepts would underpin the success of Dubai as
a top retail destination both within GCC and globally.

• Within a ten-year period, Dubai Shopping Mall GLA went from
500,000 sq m to 4,250,000 sq m, an eight-fold increase (Colliers,
2007). Yet, solid population growth and increased tourism was
seen as a key driver to this new retail development.

• In a similar light, hotel development in 2007 was being driven by
optimistic growth forecasts in tourism and the aggressive expansion of Emirates Airlines, was a key enabler to supporting new
hotel and retail development in Dubai.

Over 90% of the UAE's oil and gas reserves are Abu Dhabi based therefore one might define Dubai as a post-oil state, with its oil production peaking in the early 1990s. Despite common misconceptions, Dubai's economy
has not been heavily reliant upon the oil markets making note of its ambitions to have a more diversified economy (as seen in Vision 2021). Whilst
during the 1980s, oil revenues accounted for approximately 50% of Dubai's
GDP, in the early 2000s, this fell to 5.5%, to 2% in 2008 (Explorer, 2009) and
currently accounts for less than 1% of GDP. Nonetheless, being part of the
UAE and supportive neighbouring economies that are more heavily bound
to oil markets, Dubai's economy is still somewhat impacted by oil market
cycles. Figure 1.1 illustrates the impact oil prices has had on Dubai's real
estate (residential).

The correlation implies that the oil price declines in 2008 H2 and 2014 H2
had an adverse impact on Dubai's real estate. This was also observed with the
economic slowdown and a decrease in market sentiments in late 2019–2020
when the global price of oil fell to below $20 a barrel. Since then, the oil mar-
kets have somewhat recovered and at the time of writing are sitting around
the $120 per barrel mark (June 2022). Although Dubai's economy is com-
paratively highly diversified, falling oil prices still impact the ability of some
buyers to invest in the local residential market (KPMG, 2016). A longer-term
correlation between oil and real estate prices seems much lower.

In support of a move away from oil dependency, nation building and government investment in key infrastructure has also boosted the local real
estate market and wider economy. During the 1990s and early 2000s, Dubai
focused on becoming an international business centre. This focus was on
offering commercial organizations an attractive tax haven for business operations, development was undertaken by state-supported companies and
demand was attracted by a series of "free zones". This inherently led to

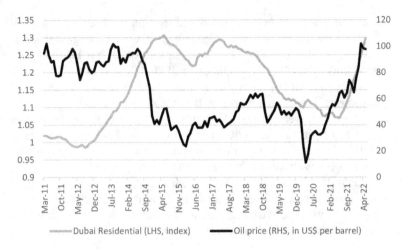

Figure 1.1 Correlation between oil price and Dubai residential real estate.
Source: Data from Bloomberg and Property Finder.

demand for commercial office space and business premises. The influx of expatriate workers to support this rapid business growth fuelled demand for residential property.

The urban population of Dubai has been rapid since 1970 and its urban fabric also extended rapidly by approximately 400 times (Ogaily, 2015). According to the Dubai 2020 masterplan, the urban population has grown based on a high growth forecast between 2010 and 2020, reaching annualized population growth of 6% per annum. Yet the city was not built like other cities, an economic response to population growth or forecasts rarely existed in the early development of Dubai. Instead, the impetus of development was much more visionary. The goal was to make Dubai the world's best city. New development would drive the economy and attract large sums of foreign investment. In the urban plans, mega projects were statements of intent rather than built from the typically planned states where supply and demand dictate. Since 1950, Dubai Municipality has commissioned a total of six urban plans (1960s, 1985, 1995, 2003, 2011, 2021). Dubai saw remarkable growth over the past 40 years in terms of its population, growing from 370,000 in 1975 to over 2 million in 2014, growing to 3.2 million in 2018 and projected to reach 5.8 million by 2040.

In summary, three key geographic and economic forces have led to the historical success of Dubai. These include:

- First, its geographical position and location have benefitted Dubai, the mid-point corridor between East and West, and located in very close proximity to traditional trade hubs such as India and the Far East. Its landmass is approximately 4,000 km² and is home to 3.2 million inhabitants (2021), relatively small in global terms yet exponentially when compared to its population of 20,000 in 1950 (represented as a 10.6% per annum population growth).

- Second, oil wealth creation from within the UAE, drove a mega-project development frenzy since its inception in 1971 (Ponzini, 2011). Fast-paced urban development from 1966 (oil discovery in Dubai) onwards gave way to capital-intensive infrastructure spending. Dubai was visioned as one of the most successful cities in the world. The infrastructure at this point had to play catch up with the trading policy that had been put in place. Dubai wanted to learn from the world's best cities and create something of a blended city – a city that would draw on replicating the successes of global cities and correcting the apparent mistakes of others.
- Third, a significant level of well-positioned trade-based (port) infrastructure. Akhavan (2020) claims that *"the entrepôt characteristics set the basis for developing an economy based on international trade, initiated with expansion of the Creek, followed by port constructions and expanding re-export activities through free trade zones."* The evolution of such key infrastructure is mapped out sequentially through the historic examination of Dubai's urban plan (see later).

Macroeconomic history of Dubai in the 21st century (2000–2021)

Historically, the oil sector may have played a role in Dubai's economic expansion, however during the two last decades, Dubai relied on business activities. In 2005, revenue from oil and gas accounted for less than 6% of the emirate's revenues. More significant contributions to Dubai's GDP in 2005 were 25% from aviation-related services; 22.6% from real estate investment and construction; and over 40% from trade and finance services (Dubai Municipality, 2014). As a result, Dubai became an early proponent of economic diversification, especially in construction and real estate. Between 2000 and 2008, the economy of Dubai multiplied four-fold reaching a GDP of $82.2 billion in 2008 (Sadik and Elbadawi, 2012). As with many other global economies, the 2008 Global Financial Crisis (GFC) saw a marked deceleration in economic growth, with a real GDP decline of 2.7% in 2009 (see Table 1.2). Tourism, finance, media, IT, healthcare, education have all emerged in the modern-day development of Dubai. Of notable importance to Dubai's economy since the early 2000s has been real estate and construction. Table 1.2 also highlights to what extent real estate has contributed to the economy of Dubai.

Real estate was undoubtedly a huge contributing factor to the retraction of economic growth, not only given the underlying global nature of the financial crisis hitting hard on the property sector, but its imperative importance as a GDP contributor in the build up to the GFC. When referring to

Table 1.2 Global % GDP change vs Dubai % GDP change

	2006	2007	2008	2009	2010	2011	2012	2013
Global	5.2	5.3	2.8	−0.6	5.3	3.9	3.2	3.4
Dubai	17.5	19.3	4.1	−2.7	2.8	3.7	3.6	4.8
% Real estate	7	8	8	7	7	6	7	7
	2014	2015	2016	2017	2018	2019	2020	2021
Global	3.4	3.2	3.2	3.8	3.7	2.8	−3.2	
Dubai	4.5	4.0	3.1	3.1	2.1	2.7	−10.7	
% Real estate	6	6	7	7	7	7	8	

Source: Dubai Statistics Centre; DLD 'Annual Report: Real Estate Sector Performance'.

Table 1.2 real estate has consistently contributed 6%–8% to Dubai's GDP. At the onset of the GFC, it was reported that residential prices in Dubai fell by 30%–40% (Sadik and Elbadawi, 2012). The diversification away from a reliance on real estate-led growth became apparent after the GFC, yet real estate development will still be a mainstay of economic activity for Dubai. Key priorities post-GFC were service-centre roles based on strong investment in aviation, shipping and logistics infrastructure (Bloch, 2010). These sectors have grown in their contribution to the Dubai economy since 2010 by approximately 40% (Dubai Statistics Centre, 2021).

The economic retraction of Dubai was boosted by the 50% growth of government spending as a fiscal response to the crisis. Key responses to the management of the GFC by the Dubai Government included:

- Fiscal government spending in infrastructure development
- Increasing liquidity in the UAE banking sector (in-line with the US)

Since the GFC in 2008/2009, there has been some economic commentary underpinning the reasons and political responses that occurred since. Renaud (2012) explored the financial crisis and real estate bubble dialogue. His analysis found that Dubai was able to prevent a much more costly financial and real estate bubble due to the quick federal response from the UAE government. This similarly appeared to be the case during its response to the economic shock of the global health crisis and COVID-19.

Dubai saw a steady recovery in economic output and growth over the next ten years. Between 2010 and 2019, the size of Dubai's economy had grown from US$80.8 billion to US$112.1 billion, a growth rate of 38.6%. A key success of Dubai's economic policy has been the establishment of free zones, such as the DIFC. Box 1.3 summarizes the growth and development of the DIFC as one successful example.

The economic conditions in 2020/2021 are like other global markets, posting uncertainty and fiscal deficits from the fallout of managing COVID-19 and the large pauses artificially placed on economic activity. Key sectors of Dubai's economy, tourism and hospitality, were perhaps hardest hit, as hotel

---

Box 1.3 In-focus: Dubai International Financial Centre (DIFC)

According to the Global Financial Centres Index, which since 2007 has ranked cities according to a range of financial, economic and quality-of-life measures, Dubai has steadily closed the gap with the top tiers of London, New York and Frankfurt. In 2020, Dubai was ranked just outside the top ten. The next highest Middle Eastern centres are ranked far behind, testament to the ability of Dubai to attract foreign talent and businesses (The Economist, 2020)

The DIFC has grown into an impressive cluster of banks, fund managers and law and accounting firms, with over 2,500 registered companies – 820 of them financial (The Economist, 2020). According to the same report, it is home to 17 of the world's top 20 banks ($180 billion assets under management). Fund managers in DIFC manage assets worth in excess of $425 billion. The key advantages of the DIFC include its attractive tax regime, and a more familiar legal system (based on English common law).

---

occupancy fell to approximately 35% during COVID-19 travel restrictions (Colliers, 2021). It would be unreasonable to expand too much on the impact of COVID as we are currently unable to assess the final impact until more time has passed. The UAE, and to some extent Dubai, is likely to feel the spillover impact of lower global oil demand owing to the global economic downtown coming out of this health crisis, most notably through restrictions of international travel and aviation (a key sector highlighted above in playing a role in Dubai's economic diversification). Economic commentators noted the weakening demand during 2020 for real estate, yet a notable price recovery since then has taken place in 2021–2022.

The final point to Dubai's economic development success has been its largely successful global marketing and branding campaign that has seen a huge emphasis placed on what Dubai can offer its residents. One notable segment of population growth has been attracting HNWIs and UHWNIs (see Box 1.4).

Box 1.4 Dubai: A home to the UHNWI

Dubai has and continues to rise in the global rankings, be it in market transparency or city competitiveness. As the city develops, so does it seem to attract more and more residents. In 2022, the UAE topped the global list of "millionaire migrants", with 4,000 new ultra-high net-worth individuals (UHNWIs) moving into the country (Ang, 2022).

The appeal of Dubai includes its relative affordability; tax-free status and a currency peg to the USD which means UHNWIs from the region and further afield have chosen Dubai as a primary or secondary residence. Furthermore, the recent revisions to immigration policy (ten-year Golden Residency for top international talent as well as property investors (>AED2 million) are also promoting further inward migration. Dubai ranks highly as a lifestyle destination; supported by a developing cultural and architectural scene; health and recreation; and a high-quality food and beverages service sector.

The mega projects of Dubai, as it was committed before 2008, had a carrying capacity of around 9.5 million inhabitants, five times that of Dubai's resident population at that time. Therefore, the intention of these "planned" projects was never to satisfy merely natural population growth, but instead to attract the wealth of foreign investors. Dubai was on an aggressive global marketing campaign and looking back, the branding campaign has worked.

According to Knight Frank (2018), Dubai ranked as the 25th largest city by the number of UHNWIs (1,060 people), with that number expected to rise 60% in the next five years (up to 2026). According to New World Wealth (2021), the total wealth held in the city amounts to US$530 billion. Dubai is the wealthiest city in the Middle East and Africa (MEA) region and the 29th wealthiest city in the world. The city is home to approximately 54,000 HNWIs and 12 billionaires. Affluent parts of Dubai include: Emirates Hills, the Jumeirah Golf Estate and the Palm Jumeirah.

Interlinkages between the macroeconomy and real estate

Macroeconomic forces are undoubtedly linked to real estate performance and one of the biggest fallouts we saw in the 2008 GFC was the significant drop in real estate asset prices both globally and in Dubai. The above data indicate that there is a relationship between economic change and real estate market behaviour. During the last 20 years in Dubai, we have seen notably property cycles between 2005 and 2008; 2013 and 2015 and an indicative upward trend emerging in 2021. Global property cycle studies indicate that although real estate cycles periodically continue, often as a result of an exogenous shock, they exhibit less volatility (Barras, 1994). The main premise behind this is the absorption of excess supply from the previous rent and capital value peaks is lagged. As a result it takes a number of years beyond a new economic growth period before new development is released, resulting often in a lower amount of new development taking place as a response/feedback mechanism to future GDP growth. Macroeconomic change does not therefore immediately result in a change in supply at a given point in time (Tiwari and White, 2010). Furthermore, global liquidity has impacted property yields (contraction) and new supply already permitted and in the development pipeline is essentially open to market forces at the time it is ready for occupation. This poses particular problems for more speculative real estate development where no identifiable occupiers are present. This might be based on economic activity for commercial occupiers or population growth/forecasts for residential occupants. The accounts of property cyclical behaviour by Barras (1994) have general applicability to all global markets including Dubai. The boom years are previously analysed as periods of large-scale real estate development, yet with any property cycle, as highlighted by Barras (1994), Dubai's story of "boom" and "bust" has not been markedly different from that of the United Kingdom in the 1980s.

Dubai's first major real estate boom from 2002 to 2008 which grew from several internal and external factors are stated below (Renaud, 2012):

- Global credit boom (2001–2008). A key feature of a real estate boom is attributed to global liquidity and its causal link to rising asset prices. Banking institutions were more willing to accept more risk as asset prices increased.
- Strong oil price performance (2002–2008), peaking at $147/barrel.
- Open risk-taking legislation and rapid infrastructure development. Dubai has outperformed the surrounding Middle East region offering firms and individuals a low tax environment and political stability. The growth strategy of Dubai centred on six key areas: trade and transportation; logistics; professional services; tourism; construction and financial services, often quoted as the Singapore or Hong Kong of the Middle East. The growth vision of Dubai has principally been built on Dubai becoming a leading global city, built off significant investment in economic infrastructure over decades (Bloch, 2010).
- Open foreign real estate ownership laws in 2002. Known as the "freehold" law, Dubai permitted foreign ownership of real estate in designated areas across the Emirate. The official announcement in 2002, Dubai saw a significant boost in real estate demand from both investors and resident expats who had the first opportunity to shift from a renting to owner occupancy model.

Despite criticisms of overdevelopment in 2008, demand for housing far exceeded supply, pushing up both capital values and rental rates (by circa +50%). This euphoric demand was matched in other real estate assets, such as hotels and offices, which also saw huge price increases. Despite the completion of some of Dubai's best-known real estate developments (Atlantis Hotel on Palm Jumeirah), there was no escaping the global real estate recession. The crisis brought a sharp collapse in demand for residential property and house prices fell sharply (by circa 25% by the last quarter of 2008) and fell by a further 25% in the following quarter.

There were a number of key responses to the crisis in 2009, most notably the $10bn bond-buying bailout made by its oil rich neighbouring Emirate, Abu Dhabi (Khalaf and Kerr, 2010), which subsequently gave some support to the falling market confidence that was embedded around Dubai's "debt crisis" at that time. Shortly after, the real estate market started to show signs of growing confidence, a measure perhaps that the worse of the GFC fallout was behind us. From late 2011 to early 2012, a steady increase in property prices began to mark the start of Dubai's second residential property cycle.

Chapters 2, 3 and 5 highlight in more detail both the residential and commercial property cycle. Later studies have argued that global markets are more synchronous given the tangible links different markets have with similar economic and financial policy. Of course, a notable observation in Dubai and the UAE is that the monetary policy and responses are steered out of the US Federal Bank. For example, as US cut interest rates in response to COVID-19, the UAE Central Bank followed. Whilst macro-level systems like this can be very similar, specific local factors and market participants' behavioural responses can be markedly different. This would lead to non-synchronous links between timings of cycles; uptake of space and levels of new development activity. Thus, a range of both local and global factors will impact the real estate cycle in Dubai.

Dubai's urban development path

Between 2002 and 2008, Dubai's population doubled and its urban footprint increased four-fold (Brook, 2013). In the mid-2000s there were contradictory views on Dubai's urban development path. It is easy for the critics of Dubai to raise up during the financial crisis or other recessions, yet I believe the vision of Dubai is in fact heading to where it has always wanted to go. The Dubai 2020 plan outlined the urban parameters of Dubai. Area 1 (offshore) and Area 2 (metropolitan/urban area) are the main focus of ongoing future development. Areas 3 and 4 are currently reserved as "non-urban", suggesting the core focus of mega-projects will sit within the Area 1 and Area 2 boundaries. This is also suggested in the recent 2040 plan, whereby Area 2 becomes the focal point of the five urban centres of Dubai (see Table 1.4).

Dubai has undoubtedly done things on a mega-scale and the achievements of urban development comprise a very long list. Dubai developed a significant amount of commercial and residential space, not based on current data but very much on economic forecasts, enabling the presence of the right amount of space for the future population of Dubai, which is now expected to reach 5.8 million by 2040. One key economic question regarding the urban development of Dubai has been "Why so high in Dubai?", despite the apparent amount of land supply available. The next section discusses this phenomenon in more detail.

"Why so high in Dubai?"

The UAE ranks fourth in the list of global countries and is home to the highest number of buildings above 150 m and 80% of those towers are in Dubai. The number of buildings in Dubai over 150 m has grown from 130 in 2011 to 232 in 2021, with 29 more under construction (Generalova and Generalov, 2018). Burj Al Arab was the Emirates' first tower to top 150 m built in 1999. Furthermore, only four buildings above 150 m were built before 2001. Over the last 20 years, Dubai has become synonymous with the rapid pace of development and an exuberance of tall buildings. A notable claim of visionary excellence. Dubai like any other city has been built out from a range of economic forces, choices of firms/producers and consumers/households to locate and occupy space (agglomeration and clustering). As competition for space increased, so did the height of the buildings. One justification might be to do with "economies of scale" and the provision of infrastructure. As the city emerged, it was more economical to have firms and households in close proximity to reduce the inefficient cost of urban sprawl. But traditional economics and bid-rent competition cannot be the only reason why Dubai has so many tall buildings. Figure 1.2 shows the historical development of tall buildings in Dubai, with a notable building boom between 2007 and 2011 (an increase of 380%).

The chart below shows that a large wave of development of tall buildings in Dubai took place between 2007 and 2010 (approximately 40% of Dubai's

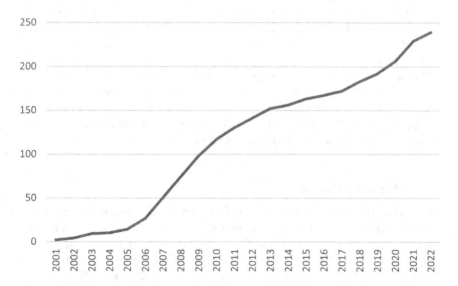

Figure 1.2 The development of tall buildings in Dubai.
Source: Council on Tall Buildings and Urban Habitat, the cumulative number of buildings > 150 m per year.

Table 1.3 Top ten tall buildings in Dubai

Building name	Location	Development status/year	Height	Land use
Burj Khalifa	Downtown	Completed/2010	828 m	Mixed
Marina 101	Marina	Completed/2017	425 m	Residential
Princess Tower	Marina	Completed/2012	413.4 m	Residential
23 Marina	Media City	Completed/2012	392.4 m	Residential
Elite Residence	Marina	Completed/2012	380.5 m	Residential
Address Boulevard	Downtown	Completed/2017	370 m	Hotel/ Residential
Ciel Tower	Dubai Marina	Under Construction/2023	365.5 m	Hotel
Almas Tower	JLT	Completed/2008	360 m	Office
Gevora Hotel	DIFC	Completed/2017	356.3 m	Hotel
II Primo Tower 1	Downtown	Completed/2022	356	Residential

Source: Council on Tall Buildings and Urban Habitat.

tall buildings were completed during this time), with some resurgence also in 2021. One might argue that the 2002 to 2007 property price growth from foreign ownership laws served as a boom for the construction of tall buildings in Dubai. Notably for a city with the world's best advertising campaign, Dubai's tall buildings might also serve as "branding", boosting the economy and attracting tourists to visit the city's awe-inspiring architecture. The multiplier effect tall buildings play in Dubai could be justifiable rationale for their existence. Of course, Burj Khalifa is the world's tallest tower at 828 m high, Dubai also tops the global list when counting the number of buildings above 300 m, totalling 27 versus 14 in New York. Table 1.3 highlights Dubai's Top ten tallest towers.

Single-use residential towers appear to dominate the list and in fact, over 50% of Dubai's tallest buildings are residential use. The trend of building high does not appear to be stopping soon, with a number of new proposed tall buildings in the height range of 350-500m, most notably the recently announced launch of Binghatti Burj in late 2022.

Dubai: The new global city?

The mega-trend of decentralization followed by globalization was born from low-cost transportation and technology, both of which impact the economics of accessibility. These forces are also increasing the opportunities for developing cities to boost trade; migration; and attract labour and capital. The choice of firms and consumers are no longer confined to national or regional boundaries. The world has become more globally competitive, and cities are now competing to attract business; tourism; skilled labour; innovation; global events and institutional investment. Measuring and comparing city performance helps understand the processes of urban change. It also fulfils a better appreciation to what political or government policy works. A significant amount of quasi-academic literature has emerged, seeking to measure the success of cities, often benchmarking or indexing the top global cities. Hence we have seen the firms such as PwC, Mercer, JLL and Knight Frank publish periodic reports on global competitiveness and city benchmarking. This competition drives cities to become more business friendly and investment ready and invites them to define their niches and specialisations more precisely. In 2022, a new type of "global city" is emerging. Dubai is globally orientated offering competition for traditional centres. Global cities have typically emerged as hubs for specific specializations, yet Dubai seemingly is offering multiple specialisms within one city. Dubai is a lifestyle capital and a significant tourism destination; financially focussed and has an emerging growth economy in knowledge and technology. The roadmap of the future of Dubai is set out in Dubai's latest plan, Dubai 2040. The final section of this chapter examines the latest of Dubai's urban development plans.

Dubai 2040 and what the future will look like?

The recently launched 2040 Master Plan maps out the next 20 years of growth and development for Dubai. The plan puts sustainability and more specifically its people at the centre of this vision. A spatial plan elsewhere might typically map out the next five years, but the long-term "Dubai business plan" can be much more useful, for both investors and developers. At its most basic level it shows us the 'where' of future city-building. The plan has selected to keep new development within its existing urban areas (Area 1 and Area 2), underpinning the need to preserve natural resources and conservation further out, keeping consistency with the previous historical plans of Dubai.

What the plan also achieves is an investor roadmap. In the next 20 years, Dubai's population is expected to grow to 5.8 million, from 3.3 million currently, putting a further catalyst to how supportive the Dubai 2040 plan will be to attracting new sources of real estate investment.

For residents, the 2040 Master Plan means improved living standards. An abundance of green space and public realm infrastructure will only enrich the high-quality of living standards Dubai already provides. The recent health crisis has placed emphasis upon people and businesses to think about social enrichment and well-being. Addressed in the 2040 plan, Dubai will see the amount of green and recreational space double; public beaches will be increased by 400% and the hospitality offerings more than doubled to keep pace with the lifestyle benefits Dubai has been accustomed too. The plan is not all about development, it also covers an equally important role

Table 1.4 The five urban centres of Dubai in 2040 plan

Urban centre	Key focus
1. Deira/Bur Dubai	Historical core centre: preservation of tradition and heritage
2. Downtown/Business Bay	Financial hub: international hub for economic, financial and business activity
3. Dubai Marina/JBR	Tourism/entertainment: leisure and hospitality
4. Expo 2020/District 2020	New core of economic growth (smart technology/ innovation); global exhibitions and events
5. Dubai Silicon Oasis	Knowledge and innovation

Source: Dubai Government (2022), access the details of the full plan via https://u.ae/en/about-the-uae/strategies-initiatives-and-awards/local-governments-strategies-and-plans/dubai-2040-urban-master-plan.

of preservation. 60% of Dubai's land mass will be allocated to preservation and nature; and social mobility and connectivity is centre-stage, particularly focussed on increasing the number of walking and cycling routes.

The clarity of the plan's five urban centres builds confidence in Dubai becoming a city of complementing polycentric development rather than competition. This will undoubtedly boost the economic growth and development required to meet the future population aspirations. As noted in the earlier parts of this chapter, the plan also serves a roadmap of future economic diversification, advancing the established business sectors whilst also paving way for many new sectors. Table 1.4 shows the five urban centres of Dubai from which new development and redevelopment is expected to be focused. Key outcomes of the Dubai 2040 urban masterplan include:

- Total resident population planned to be 5.8 million by 2040
- Addition of two new main urban centres
- Increase of green and recreational space by 105%
- Increase accessibility to main public transport stations (55% of population within 800 million)
- Increase space provision of health and education facilities (+25%)
- Increase space provision for hospitality and tourism (+134%)
- Preservation of nature and conservation (60% rural area)

Towering ambitions: Dubai's polycentric growth

Housing types and affordability are spread across the city and are not so correlated to the bid-rent curve assumptions made in a hub and spoke city. Dubai like many other global cities has developed away from one central focus, instead developing along a polycentric model. Notably, these urban nodes are characterized by the presence of tall buildings, an indication of where economic forces have driven up land competition and higher-density development. Prior to modern-day development and the establishment of free zones, Deira was the monocentric focus of Dubai, built off its location as a trading hub and developed around its waterways. Each urban node in Dubai is an economic powerhouse, often known for a particular component of Dubai's economic fabric. For instance, Dubai Marina serves a touristic focus; and Downtown/DIFC are financial centres. New centres like Business Bay and JLT are perhaps sprawling off the success and close proximity to Downtown/DIFC and Dubai Marina, respectively. Dubai Creek Harbour represents a new urban node with its own economic impetus (see Box 1.5).

Box 1.5 Dubai Creek Harbour – A new part of Dubai's polycentric development

- Dubai Creek Harbour is one of Dubai's newest large-scale urban masterplans, developed by Emaar, offering a similar schematic design to Dubai Marina some 30km away. In 2016, Dubai Creek tower was announced as part of the development, and was planned to be taller than the Burj Khalifa some 10km away. According to Emaar, the development is planned to be home to 712,000 sq.m of serviced apartments; 7.3 million sq.m of residential space and 700,000 sq.m. of open parks and green space. Many of the current residential buildings in Dubai Creek Harbour are above 50 floors, showcasing that Dubai continues to have an appetite to build tall towers.
- Dubai is a clear global example of a polycentric urban development form – from Downtown Dubai to the Marina; from Dubai Creek to Deira, few of the many districts that form the historical development of the Emirate. Within the largest centres, we see tall buildings and high density (Safarik et.al., 2018).
- Deira, was Dubai's original centre, built around its water networks.
- Dubai International Airport is now the epicentre of Dubai's urban economic development.
- Vertical "tall building" development separates the urban clusters and offers a visual barometer of economic success within the city. Architectural beacons separate geographical space. Horizontal patterns are more restricted as mapped in Figure 1.3.
- Dubai is also seeing some transportation innovations, from autonomous driverless vehicles to the Hyperloop project that might see an urban reconfiguration in the access-space proposition.

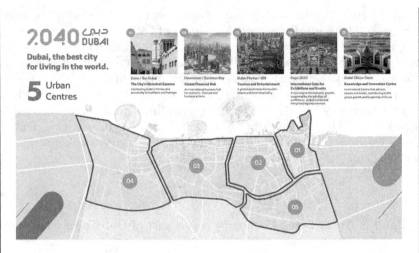

Figure 1.3 Image of the Dubai 2040 Masterplan.
Source: Provided with permission from RTA

Box 1.6 Expo 2020/District 2020: Building the legacy

Expo 2020 in Dubai was the first of its kind in the MENA region. For many cities, hosting a mega-event is now the single largest undertaking in urban development. One of the key attractions of hosting mega-events is the promise of a catalytic effect on urban development "accelerating infrastructural development by up to 10 years" (Preuss, 2004). In the case of Expo 2020 in Dubai, the event has been a catalyst for new development – major infrastructure means that the city expands; new business sectors emerge and a supportive range of residential; commercial and leisure amenities are introduced into the city. Similarly, expenditure on hosting a mega-event is often justified via the multiplier effect and a boost to the national economy. Ernst and Young reported that hosting the event will boost the economy by as much as AED122.6 billion ($33.4 billion) (The National, 2019).

Why was the Expo 2020 located where it was?

Located close to Dubai World Central airport in Dubai South, the site of the event is expected to cover a total of 438 ha (Oxford Business Group, 2020). The new growth district of Expo 2020 is equidistant between Abu Dhabi International Airport and Dubai International

Airport (DXB), as well as being within a 15-minute road transportation from Jebel Ali Port (the world's ninth largest container port). District 2020 as a new free zone will offer incentives such as:

* 0% corporate and personal tax
* 100% foreign ownership and repatriation of capital
* Office space options from low-rise campus style buildings to grade A high-rise buildings

The location within a host city matters. According to Sroka (2021), a high land value location can prevent prospective spillover activity from private sector developers (who are "priced out") and mean the need for a greater government spend and conversely, a lower-value location can unlock or capture upside without the considerable real estate costs of an already high demand/highly priced location Expo 2020 is identified as a new urban centre for Dubai and so in fact hosting the event has provided impetus to the development of necessary infrastructure, such as roads, bridges and the extension of the Dubai Metro Red Line, Route 2020. This infrastructure spend has then led to a significant amount of capital being invested on construction projects related to Expo 2020.

What are the benefits of hosting a mega-event?

The first supportive argument for hosting a global event would be the potential increase in visitor numbers, a spotlight or halo effect that brings a new attention and vibrancy to the host city. Similarly to the economic multiplier effect of government stimulus packages in relation to a national economy, expenditure on hosting a mega-event is often justified via the multiplier effect. A principle that states an initial impetus is paid back magnitudes more through a range of supporting expenditure patterns (for instance, more visitors; better hotel occupancy; more consumption in leisure and retail, leading to more employment and more jobs and higher propensities to consume further down the line). Though critics have highlighted that economic studies have often applied incorrect multipliers and a range of historic studies have found little economic impact on hosting major events. Real estate studies in a range of global cities do find a positive spillover effect on property values. The results of Coates and Matheson's (2011) analysis indicate that rental values are generally affected by hosting these mega-events but not in a consistent manner. Though, larger-profile global events like Olympics and World Cup did show more significant results.

Muller and Gaffney (2018) research found that cities with more market-led economies were better able to use the event for urban development, explaining that major host events reconfigured urban governance arrangements and strategic development plans. We have seen similar in Dubai with the strategic development importance of District 2020 (post-Expo site) within the Dubai 2040 Urban Plan.

One key term used by Muller and Gaffney (2018) was "entrepreneurial urbanism" which found most host cities benefited from an entrepreneurial rhetoric very prominent in the preparation for the mega-event (perhaps building upon "Brand Dubai"). District 2020, the legacy component of Expo 2020 is a well-thought-out proponent of this and the event organizers have planned carefully about how the buildings and land uses will evolve post-event. A key part of this has been a leasing up strategy to like-minded private organizations who see what will be left behind after Expo 2020. District 2020 will thrive as an entrepreneurial hub for years to come. Dubai is likely to be somewhat more bullish about the Expo 2020 city-branding opportunity, as there are still aspirations for Dubai to become a leading global city, and Expo provides this framed around a future-looking arena, showcasing Dubai as a modern, technologically advanced and innovation-driven city.

Hosting Expo brings opportunities for future commercial enterprises. Vancouver hosted the Expo in 1986 and then later hosted a range of much more prolific mega-events, this event leveraging could also be positive for the longer term in Dubai or wider UAE. Certain similarities can be drawn with the convention infrastructure being a key part to Dubai's Expo legacy planning. Akin to London and the creation of the Olympic Park, Expo 2020/District 2020 will become a nucleus for new property development. As it is identified as one of five new urban centres in the Dubai 2040 masterplan, this process has already somewhat begun. District 2020 will house the largest DWTC Exhibition Hall and Conference Centre which relocates this away from its existing Trade Centre/DIFC location. For cities, one of the key attractions of hosting mega-events is the promise of a catalytic or spillover effect. Other studies have have found that hosting international events is an opportunity to reimagine or re-plan a city (Kassens-Noor, 2012), as might be the case with the Dubai 2040 masterplan. In Vancouver and London, hosting mega-events helped confirm existing plans for urban development and made available the government spend to invest in bringing such plans to life. Notably the success of the London Olympics in 2012 was the fact it was part of an existing redevelopment program in East London and financial government support was made available to accelerate these plans (Poynter, 2009). District 2020 is based very much on a similar grounding post-Expo-2020.

Beyond the Expo 2020 (October 2021 to March 2022), 80% of the built environment will be repurposed in District 2020 (from October 2022). Key metrics within the new urban centre will be:

- 55 LEED-certified commercial buildings (minimum Gold)
- 700 new residential units
- 3,000 new hotel rooms
- 108 serviced land plots

The district will also be exemplar in terms of Smart Mobility, utilizing 5G technology throughout and is set out to be a fully integrated human-centric smart city, bringing benefits of a more balanced and sustainable lifestyle for residents and businesses. District 2020 will be the region's first WELL Community Standard development bringing a range of health and wellbeing standards to form part of the masterplan. Key metrics within this standard include:

- Green spaces: Parks and green spaces provide benefits for both mental and physical health
- Wellness: A walkable realm and network of cycle paths for daily commuting; physical activity, reducing reliance on carbon-emitting vehicular activity
- Comfort: A range of shading and natural ventilation in design and new innovative cooling technologies to promote comfort

What is the impact on property prices?

Real estate studies in a range of global cities do find a positive effect on property values. Though as we saw these property price increases are not simply linear. The residential market expected a boost based on a significant number of international visitors coming to Dubai for the Expo 2020. Off-plan developments near the Expo 2020 site are attracting both buyers and investors, particularly the units in Dubai South – where rental yields in properties reached 11% in early 2019 (Oxford Business Group, 2020). Roche (2021) published a discussion paper on the impact on real estate for the previous World Expo hosts. They concurred that hosting an Expo event had a largely positive impact on house prices, somewhat more pronounced price growth in the five years post-Expo (see Table 1.5).

Table 1.5 The impact of hosting Expo v House price growth pre-/post-event

Expo host		House prices, 5 yr pre-Expo (%)	House prices, 5 yr post-Expo (%)
1992	Seville, Spain	+23	+15
1998	Lisbon, Portugal	+18	+60
2000	Hannover, Germany	+12	+40
2015	Italy, Milan	−2.5	+40
2020	Dubai, UAE	+2.9[a]	n/a

Source: Roche (2021).
a Author extracted and added Dubai data from DXBInteract.com.

Table 1.5 shows that the largest price increases in hosting the mega-event takes place in the five years post-event. So are we on a similar trajectory in Dubai post-Expo? Real estate transactions and prices were both heavily impacted by the the COVID-19 pandemic. However, residential prices recovered strongly during the run-up to the event in late 2021, especially the high-end villa market, as real estate buyers worldwide began "a race for space". Since then, Dubai in 2022 has been observing a post-Expo residential property boom with record-breaking transactional values and sales volumes. A market report by Property Monitor shows that property values in Dubai now stand at AED1,065 per square foot, a level last seen during Dubai's previous residential up-cycle in 2013. Additionally, it is reported that trans-action volumes in October 2022 stood at 8,626, a growth trend that has been steady throughout the year. Furthermore, it is expected that 2022 will see approximately 90,000 residential sales transactions re-corded, its second highest level since 2002 (Property Monitor, 2022). Post-expo, Dubai has already witnessed a significant price boost in its average property prices, rising 7.5% since the start of 2022.

Another new growth area presented in the plan is District 2020/Expo City, the legacy project of Expo 2020. Box 1.6 reviews the impact of Expo 2020 as a mega-event and beyond.

Conclusions

Dubai is a fascinating global real estate development case study. The tow-ering ambitions and scale of development drew a spotlight on what Dubai has become to a global audience. Whilst at face value critics would make calls that it has appeared to have ignored the economic parameters of de-mand and supply, through my analysis, I believe it has in fact been working on something much bigger. Dubai over the last 50 years has been built and planned to become the best; a leading top-tier global city, drawing on both international best practices and aligning to its own strategic vision. Real estate development has consistently been a significant part of Dubai's GDP and boosted the economy away from any over-reliance the UAE may have had historically on oil revenues. Ambitions to be the best have always been a top priority for Dubai and good long-term governance has been a key driver to its success.

After discussion, this chapter has provided a summary of these processes and looked at what might be expected in the next 20 years. It is hoped that it provides a robust overview of how Dubai took advantage of its geographical trade position in its early development years, and how it has been agile and advantageous to the new opportunities presented in our modern global economic era. The background to Dubai presented in this chapter can be carried forward into the remaining parts of this book and gives context to a better understanding of the evolution of Dubai's real estate market.

References

Akhavan, M. (2020) 'Chapter 4 making of a global port-city in the Middle East: The Dubai Model', *Port Geography and Hinterland Development Dynamics: Insights from Major Port-cities of the Middle East*, Singapore: Springer, pp. 51–69.

AlShafieei, S. (1997) The Spatial Implications of Urban Land Policies in Dubai City. Dubai Municipality: Unpublished Report cited in Akhavan, M. (2017) 'Development dynamics of port-cities interface in the Arab Middle Eastern world – The case of Dubai global hub port-city', *Cities*, 60, pp. 343–352.

Ang, C. (2022) Mapping the Migration of Millionaires. https://www.visualcapitalist.com/migration-of-millionaires-worldwide-2022/.

Barras, R. (1994) 'Property and the economic cycle: Building cycles revisited', *Journal of Property Research*, 11 (3), pp. 183–197.

Bloch, R. (2010) 'Dubai's long goodbye', *International Journal of Urban and Regional Research*, 34 (4), pp. 943–951.

Brook, D. (2013) How Dubai Became Dubai. http://nextcity.org/daily/entry/how-dubai-became-dubai.

Coates, D., and Matheson, V.A. (2011) 'Mega-events and housing costs: Raising the rent while raising the roof?' *The Annals of Regional Science*, 46, pp. 119–137.

Colliers (2007) Dubai Real Estate Overview: Market Research, Q4 2007 (printed). Dubai: Colliers International.

Colliers (2021) MENA Hotel Forecast, April 2021. https://www.colliers.com/en-ae/research/dubai/mena-hotel-forecasts-april-2021.

Council on Tall Buildings and Urban Habitat (CTBUH). https://www.skyscraper-center.com/city/dubai (Accessed September 2021).

Dubai Government (2022) https://u.ae/en/about-the-uae/strategies-initiatives-and-awards/local-governments-strategies-and-plans/dubai-2040-urban-master-plan.

Dubai Municipality (2014) https://isocarp.org/app/uploads/2014/05/AfE_2012_-_Dubai_Municipality-_Planning_Department.pdf?cv=1

Dubai Statistics Centre (2021) https://www.dsc.gov.ae/en-us/Pages/default.aspx.

Elbadawi, I.A., and Soto, R. (2012). 'Sources of economic growth and development strategy in Dubai', In: Sadik, A.T.A., Elbadawi, I.A. (eds) *The Global Economic Crisis and Consequences for Development Strategy in Dubai. The Economics of the Middle East*, New York: Palgrave Macmillan, pp. 121–153.

Explorer Group Ltd (2009) *Dubai Complete Residents' Guide.* (13th Edition). Dubai: Explorer Publishing.

Generalova, E.M., and Generalov, V.P. (2018) 'Residential high-rises in Dubai: Typologies, tendencies and development prospects', *Council on Tall Buildings and Urban Habitat Journal*, Special 2018 Conference Themed Issue: Polycentric Cities, pp. 36–43.

JLL (2020) *City Momentum Index (CMI)*. https://www.us.jll.com/en/trends-and-insights/research/city-momentum-index-2020 (Accessed 2021).

Jones, C. (2013) *Office Markets and Public Policy*, 1st Edition, London: Wiley-Blackwell.

Kassens-Noor, E. (2012) *Planning Olympic Legacies: Transport Dreams and Urban Realities*. Oxford: Routledge.

Khalaf, R., and Kerr, S. (2010) Dubai: A Trade to Ply. Financial Times. https://www.ft.com/content/48c8c9fe-879a-11df-9f37-00144feabdc0 (Accessed January 2022).

Knight Frank (2018) The Wealth Report City Series: Dubai Edition – 2018. https://www.knightfrank.com/research/report-library/the-wealth-report-city-series-dubai-edition-2018-5332.aspx.

Knight Frank (2021) Dubai Timelapse. https://youtu.be/voN9imXg6xM (Accessed April 2022).

KPMG (2016) Building Confidence: A Review of Dubai's Residential Real Estate Market. https://assets.kpmg/content/dam/kpmg/pdf/2016/04/building-confidence-uae.pdf.

Muller, M., and Gaffney, C. (2018) 'Comparing the urban impacts of the FIFA world cup and Olympic games from 2010 to 2016', *Journal of Sport and Social Issues*, 42 (4), pp. 247–269.

New World Wealth (2021) https://newworldwealth.com/reports/f/the-wealthiest-cities-in-the-world-2021 (Accessed 2022).

Ogaily, A. (2015) *Urban Planning in Dubai; Cultural & Human Scale Context*, Dubai: Council on Tall Buildings & Urban Habitat.

Oxford Business Group (2020) https://oxfordbusinessgroup.com/analysis/positive-signs-investment-property-related-expo-2020-begins-bear-fruit.

Pacione, M. (2005) 'Dubai', *Cities*, 22 (3), pp. 255–265.

Ponzini, D. (2011) 'Large scale development projects and star architecture in the absence of democratic politics: The case of Abu Dhabi, UAE', *Cities*, 28 (3), pp. 251–259.

Poynter, G. (2009) 'The 2012 Olympic games and the reshaping of East London', In Imrie, R., Lees, L., and Raco, M. (eds), *Regenerating London: Governance, Sustainability and Community in a Global City*, London: Taylor and Francis, pp. 132–148.

Preuss, H. (2004) *The Economics of Staging the Olympics: A Comparison of the Games 1972–2008*, Cheltenham: Edward Elgar.

Property Monitor (2022) 'Monthly Market Report: October 2022'. https://propertymonitor.com/insights/monthly-market-report/monthly-market-report-october-2022

Ramos, S.J. (2010) *Dubai Amplified: The Engineering of a Port Geography.* Surrey: Ashgate Publishing.

Renaud, B. (2012) 'Real estate bubble and financial crisis in Dubai: Dynamics and policy responses', *Journal of Real Estate Literature,* 20 (1), pp. 51–77.

Roche, J. (2021) Lessons from the Past, Pointers to the Future. What Impact do Expos Have on Real Estate Markets? Cavendish Maxwell. https://cavendishmaxwell.com/resources/uploads/2021/11/CavendishMaxwell.Lessons-from-the-past-pointers-to-the-future.-What-impact-do-Expos-have-on-real-estate-markets.Op-Ed.Nov2021.pdf (Accessed March 2022).

Sadik, A.T.A., and Elbadawi, I.A. (eds) *The Global Economic Crisis and Consequences for Development Strategy in Dubai. The Economics of the Middle East.* New York: Palgrave Macmillan.

Safarik, D., Ursini, S., and Wood, A. (2018) The Tall, Poycentric City: Dubai and the Future of Vertical Urbanism, *The Council on Tall Buildings and Urban Habitat (CTBUH) Journal,* 2018 (4), pp. 20–29.

Sroka, R. (2021) 'Mega-events and rapid transit: Evaluating the Canada line 10 years after Vancouver 2010', *Public Works Management & Policy,* 26 (3), pp. 220–238.

The Economist (2009) The Outstretched Palm. https://www.economist.com/finance-and-economics/2009/02/26/the-outstretched-palm.

The Economist (2020) Can Dubai Enter the Premier League of Financial Centres? https://www.economist.com/finance-and-economics/2020/08/22/can-dubai-enter-the-premier-league-of-financial-centres.

The National (2019) UAE Expected to Gain Dh122.6bn after Hosting Expo 2020. https://www.thenationalnews.com/business/economy/uae-expected-to-gain-dh122-6bn-after-hosting-expo-2020-ernst-young-finds-1.849286.

Thompson, P. (2016) Dubai: An exemplar of state capitalism, *Turkish Policy Quarterly,* 15 (2), pp. 159–169. http://turkishpolicy.com/files/articlepdf/dubai-an-exemplar-of-state-capitalism_en_6676.pdf (Accessed June 2020).

Tiwari, P., and White, M. (2010) *International Real Estate Economics,* London: Palgrave Macmillan.

2 Property market activity

The primary aim of this chapter is to provide an overview of investment markets in Dubai. The chapter introduces investing in real estate generally and provides a knowledge base for the material developed in later chapters (Chapters 3–5). It begins by considering the rationale for property investment and the qualities real estate provides an investor. It then moves on to examine what institutional funds are looking for when investing in real estate and provides market commentary on why Dubai has historically lacked the levels of institutional investment seen in other global hubs. Within the analysis, there is also an overview provided on the alternative investment products available in Dubai for both local and foreign investors to gain exposure to the Dubai property market. Given the dominance of residential investment in Dubai's property market, the chapter concludes and provides an appraisal basis from which a residential investor (taking the view of an expat) could take when looking to buy property in Dubai and what key considerations should be taken into account when making a decision to buy in Dubai.

Real estate – The global context

Real estate is the world's largest asset class, estimated to be worth over $326.5 trillion globally (Savills, 2021), which is three times larger than the global stock market and four times higher than the global GDP. A staggering 80% of this wealth is contributed from residential real estate. Yet investing in real estate differs somewhat from mainstream assets, like equities and bonds and anyone investing in real estate needs to appreciate the attributes of real estate. Box 2.1 outlines what is real estate.

Box 2.1 What is real estate? The definition

A market typically comprises the following:

- A good/commodity to be bought and sold
- Willing sellers; willing buyers
- A means of bringing buyer and seller together

DOI: 10.1201/9781003186908-3

Property markets are somewhat different to a "perfect" market, based upon:

- Land and buildings are immobile; supply is relatively inelastic; heterogenous (no two properties are identical either from a physical or legal perspective)
- No central marketplace; lack transparency; highly secret and confidential
- Price is the product of negotiation between the buyer and seller (often privately) This factor and infrequent trading makes valuations of market prices difficult.
- High transaction costs. The high cost of buying and selling property rights inhibits market participants from responding rapidly to changing conditions and increases the gross return required by investors.

The real estate market is multidimensional and complex. The first issue is with information flows and transparency, often making it hard for rational decisions to be made. It is not well-defined nor flush with plentiful amounts of comparable data to base decisions. Despite living in a world of open data and the internet, property data is still largely informal and vaguely defined.

The second issue is that when we buy real estate, we are buying legal rights, these rights then translate to rights to occupy or rights to receive an income from property. Legal rights (and not so much the physical ones) constitute value (see Chapter 9). As these rights are durable, they exist over long-time periods. Valuers even refer this as perpetuity (a measure of rights to receive income infinitely in the case of freehold ownership). At the same time, its durability and its tangible nature means real estate also depreciates and suffers obsolescence. When measuring the attractiveness of real estate, we must also be mindful of operating and capital expenditure that is required periodically over its holding period. As a final point, numerous legal rights may exist over one property at any given time (own; lease; sub-lease). Therefore, the existence of different rights segments the property market into subsections. Three discrete but overlapping sectors can clearly be identified when analysing real estate markets: the occupier/user; the investor and the developer.

Risk management in a property portfolio (direct) suffers from an inability to diversify risk, as the capital sums involved are significant (unlike an equities portfolio). The use of leverage permits somewhat a better ability to diversify but carries risk of sudden changes to market conditions and over-exposure. Gearing (leverage) can distort the risk and return profile of real estate investing. Real estate portfolios also typically carry a high degree of specific risk (non-diversifiable risk).

In terms of its investment qualities high net-worth individuals (HNWIs) and institutional investors have consistently favoured real estate in their asset allocations and noticeably shifted to higher allocations in recent years. There are many reasons why commercial and institutional residential real estate are attractive investments. Most notable these include:

- Real estate remains a good diversifier for portfolios dominated by equities and bonds and has historically been attractive, often boosting the return profile of a mixed asset portfolio
- Real estate performs well relative to other investment categories (risk/return adjusted basis)
- Real estate provides a secure and stable cash flow (via the lease)
- Real estate has offered low volatility of returns (subject to reversion to market rents, via lease length)
- Rent is a contractual obligation, dividends from the stock market are not

Key characteristics of Dubai's real estate market

One of the most common criticisms of Dubai's property market is that it is much less transparent than other global markets (US, UK, Australia) making use of the comparative method and sourcing key property information like yields rather challenging for property valuers and investors alike. Chapter 8 will discuss this in more detail. The asset characteristics of real estate that impacts the attractiveness of it as an investment asset class include:

- For real estate, **cash flow is king,** and this is delivered via the lease committed between landlord and tenant. In some markets, leases can be significantly longer than others, and more importantly, the reversion of rent to market rent can happen quicker or slower depending on market practices. UK/European commercial leases are typically 7–10 years in length. Those in Dubai average 3–5 years. Within these lease agreements, there will be specific mechanisms for reverting the rent throughout the agreement that provides income security for the building owner (see Chapter 6 for current leasing and letting practices in Dubai).
- In Dubai, property investors are investing sums in AED which is pegged to US Dollars, a widely accepted safe-haven global currency. Investing in a USD-denominated currency offers investors investment in a bellwether currency (compared to some other emerging economies).

- Rents in Dubai are often received annually in advance (or quarterly in advance). Therefore, investors receive a further boost to their real returns when compared to a monthly rental return.

- The traditional view on real estate is **supply is inelastic**. This makes real estate less responsive to demand and supply imbalances. In a bull market, new supply is limited and price escalation will occur in response to increase competition for space. In a bear market, there will be a heightened vacancy rate that typically follows voids and vacancies in non-prime assets. Any new supply will be lagged behind the demand driver, as it takes a considerable time to develop and construct new real estate assets.

- Government intervention is much more laissez-faire in Dubai meaning supply is somewhat more elastic as it often does not take as long for property in Dubai to be built. As a point of reference, the average Dubai tower takes 30 months to build. This enables supply to adjust more readily to changes in demand. Conversely, it does expose the market to periods of oversupply if property is developed in large numbers simultaneously.

- **Real estate valuations impact market behaviours.** As property is infrequently traded, valuations serve as a proxy for market information on price performance. These are imperfect signals that influence behaviour of market participants. In opaque markets, it may be argued that the price signals from valuations are less efficient than others. This impacts investor confidence.

- **Real estate is illiquid** (depending on market conditions/timing in the cycle) which can lead to periods of price volatility. Whilst it may take 3–6 months to buy and sell a property in most markets, in Dubai, a property transaction can be completed much quicker (14 days to 6 weeks is typical) depending on whether the transaction is cash or mortgage-based. Therefore, the time that it takes to receive cash tied up in property is typically much shorter when investing in Dubai. The ability to "trade" real estate is still impeded by the high transaction costs, which are comparable to other global markets (see Chapter 3).

- The ability for **real estate to be an inflation-hedge** is once again tied to the lease terms and how future rent is decided upon at a future date. If, as in some global markets, future rent is decided upon via an inflation-linked measure (RPI or CPI), then the asset is somewhat protected in inflationary periods. However, if market practices are based on a rental index, like in Dubai, it may or may not be inflationary, therefore is exposed to market/economic risk at that point in time rent is agreed. It therefore exposes assets to a similar form of equity-type risk of the real economy. A second component of its inflation hedge is more straightforward, as construction costs increase (with inflation), building replacement costs also increase. Real estate values would tend to increase in line with replacement cost overtime, therefore real estate becomes a store of wealth to some extent.

- **Real estate is a depreciating asset**, it wears out over time, and becomes both physically and economically obsolete (and perhaps nowadays environmentally). Dubai is a relatively new city, but with 44% of the residential supply older than ten years and the average age of an office building in the established commercial freezones being 15 years, depreciation will become more apparent. The Dubai Building Code (2021) refers to a recommended minimum design life of a building to be 50 years. Therefore, building owners will need to be more considered around managing depreciation and capital expenditure moving forward. One thing that may boost this separation further is the proposed rental index (November 2022) where rental increases will be bound against the performance of individual buildings rather than sub-areas or residential districts. Dubai's star rating system may permit rental increases to be more transparent and in line with the property's level of service and asset management. This is akin to the other parts of the world, where consumers and investor will start to segregate the market even further from sub-market to individual building performance.The rate of this depreciation depends on the asset class as well as the pace of innovation around the real estate space. For instance, if the pace of technology is rapid then older IT capabilities within the building will make the existing stock less attractive to occupiers than new buildings. Depreciation may also arise simply from "wear and tear" and the deterioration of the building's material component parts. Depreciation is often factored into the valuation and appraisal process (see Chapter 9).
- **Volatility** and (returns per unit of risk) are attractive for real estate assets globally, outperforming bonds and equities over medium-term periods (c. 20 years). In Dubai, the market has become somewhat synonymous with volatility, perhaps a by-product of the above reference to exposure to equity-type risk in the real economy. On a similar note, real estate returns are typically controlled by **property cycles**, which typically run 7–9 years in Dubai (see Chapter 3).

These characteristics will impact the way in which real estate is viewed in comparison to competing for alternative assets, typically equities (stocks) and bonds. A discussion on real estate pricing is explained in more detail later in this chapter (see "How is commercial real estate priced?").

Current property market performance in Middle East and Dubai

Growth in residential prices accounted for much of the annual uplift in global real estate values in 2020 (having seen an 8% rise). According to Savills (2021), residential value makes up a significant 80% of the global real estate wealth ($258.5 trillion), yet the spatial distribution of this housing wealth is uneven globally and 75% of the $258.5 trillion sits within ten countries (China; US; Japan; Germany; UK; France; South Korea; Canada; Italy and Australia). As a result, other global regions like the Middle East make up a very small proportion of global housing wealth. In a similar light, the global commercial property market is more modest at \$32.6 trillion, with half of the commercial property wealth being located in the US (27%); China (16%) and Japan (6%). In value terms, the estimated size of Dubai's commercial property transactions between 2010 and 2020 is significantly lower than other major global hub cities. According to Real Capital Analytics (2021), Dubai's value of commercial property investment in the last ten years reached \$8.45 billion, c. 6% of that reached in Singapore; 2.8% of Hong Kong and a mere 2.3% when compared to London.

Within Dubai, investment in residential property has consistently overshadowed the commercial sector in terms of investment sums and volume. Figure 2.1 provides a snapshot of real estate transactions in Dubai over the last ten years. The top part of the chart shows that commercial property investment has maintained a low range between US\$5 billion and 10 billion per annum, whereas residential transactions are magnitudes higher, ranging from a low of US\$25bn in Q1 2020 to US\$65 billion in Q3 2013. Furthermore, transaction volumes are 5–8 times higher. The next section looks to elaborate on these trends further.

Global investors especially those of a fund will typically be seeking to invest in specific types of assets. Whilst a spectrum of asset types exists, international funds will typically want exposure to Core investment-grade assets. Table 2.1 shows the main investment characteristics of Core assets in mature international markets and makes the comparison to such asset availability in Dubai. This provides a framework from which one can evaluate how attractive is the Dubai real estate proposition and what types of

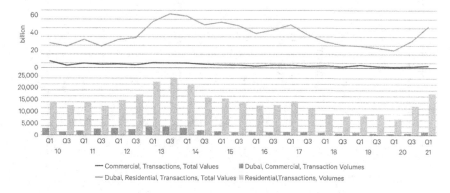

Figure 2.1 Dubai real estate transactions: Timeseries of commercial and residential values and volumes, 2010–2021.
Source: supplied by Taimur Khan.

Table 2.1 A comparison of "core" international fund requirements versus Dubai commercial property investment

Key criteria	Core (International)	Commercial (Dubai)	Other investment (Dubai) – Value-Add and New Development
Return (% p.a.) Return profile Risk Strategy	4-6% Income Low Long-term cashflow and capital preservation	6-9% (but few assets exist) Income + Appreciation Low-Mid Exposure to new markets; diversify away from traditional core markets	>15%, risk + new development Income + Value-add (development) Mid-High (as new development most likely) Buy under-performing assets, requires management to get to Grade A specification. New development of Grade A most likely option rather than redevelopment of buildings (<20 years old)
Investment/asset qualities	Prime/Grade A, long-lease, inflation-linked, excellent tenant ratings ("institutional")	Few assets on acquisition are prime, long-lease, rental increases are fixed (commercial) or follow rent index. Strata-law (single floor ownership) is unappealing. Land law legislation/title restrictions (onshore v offshore) brings higher risk. Improving underlying security of covenant and occupation	Lease management and improvement in tenant quality Fit specifics of the institution, who failed to find suitable existing stock in market.
Examples		Good examples of institutional grade assets within Downtown Dubai; Dubai Media City Dubai Internet City and TECOM freezones. Specific examples provided in Chapter 5	Brookfield Asset Management JV with Meraas (Merex) including areas of City Walk; The Beach; La Mer (Value-Add) ICD Brookfield Place (DIFC); Ikea Warehouse (Dubai South), opted for new development over repurposing existing stock.

Source: Author

assets are, and will need to be present, in the market to capture the interest of international property funds.

Box 2.2 expands on the key reasons why Dubai's commercial property sector has fallen behind in terms of global comparable cities and draws out why Dubai's residential sector has overshadowed its commercial investment sector.

Box 2.2 Industry insight by Taimur Khan, Partner – Head of Middle East Research, CBRE

Institutional investment in Dubai

Despite the rapid growth in Dubai's built environment and a corresponding increase in the value and volume of residential transactions, both from local and international investors, its commercial investment market has not followed suit. This is even more so the case for Dubai's institutional investment market, which has seen, at best, scattered activity over the last decade (see Table 2.2).

Table 2.2 Key investment transactions in Dubai (by Asset and by Sector): 2011–2021

Asset	Sector
U-Bora Tower	Office
Emaar Hospitality Portfolio	Hospitality
ADCB HQ	Office
Souq Extra, DSO Phase 1	Retail
Lycée Français Jean Mermoz	Education
North London Collegiate School	Education
The Edge Building	Office
European Business Centre	Office
Hartland International School	Education
Binghatti Vista and Binghatti Sapphires	Residential
The Ritz (Whole building)	Hospitality/Residential
The Shangri-La (Whole building)	Hospitality/Residential
Arabian Oryx House	Residential
Binghatti Terraces	Residential
Remraam Residential	Residential
The Pad	Residential
City Walk Residential Building	Residential

Source: Supplied by Taimur Khan.

This is a stark contrast to commercial real estate markets in other global hub cities. In the ten years to 2021, in key global hub cities such as those listed below, investment into these cities' commercial real estate markets, both local and cross-border, far outstrip the level seen in Dubai (see Table 2.3).

Table 2.3 Key global hub cities, international commercial real estate investment volumes

Market	Commercial real estate market investment, total value (US$)
London	$367.3 billion
Hong Kong	$297.01 billion
Singapore	$141.2 billion
New York	$107.12 billion
Dubai	$8.45 billion

Source: Supplied by Taimur Khan, Real Capital Analytics.

This gulf in total transaction value, in Dubai, is not due to a lack of domestic or regional capital either. Both public and private institutional investors from GCC countries are known to be major buyers of commercial real estate internationally. According to data from Real Capital Analytics, over the decade to 2021, almost $30 billion dollars of investment into real estate markets internationally (excluding Dubai) has originated from GCC countries. This figure is likely to be materially lower than the reality, as a result of undercounting of investments from the region due to the types of investment structure used.

Why had Dubai historically seen a lack of interest in its commercial property?

First, whilst there is a lack of institutional investment in commercial real estate in Dubai, it is important to note that this does not mean there is a severe lack of activity in the market. In fact, there has been material activity over the last decade; however, it has originated largely from individual private investors.

According to data from REIDIN, as at H1 2021, commercial real estate (excluding land) accounted for just 6.7% and 3.4% of total market transactions and value, compared to 12.1% and 13.0% a decade earlier, respectively. In reality, the height of commercial market activity in the early part of the last decade was driven by the sale of strata-titled office stock, rather than true institutional commercial investment. Much of this investment was either for owner-occupation or as speculative investment, with many developers content to sell strata titled stock, particularly as this helped fund development activity. As a result of this, currently, it is estimated that 52.8% of office stock in Dubai, out of a total of 10.43 million sq m, is strata titled.

Whilst strata-titled stock will deter some investment from institutional investors, evidence from major commercial real estate investment markets around the world suggests that there are still a number of institutional buyers who will acquire such assets. However, in Dubai, the issue around the majority of this stock lies around the quality of available stock. The latest data show that 59.5% of the strata-titled stock is of Grade B or lower quality. These factors combined equate to an effective, non-workable investment case for the vast majority of institutional investors and one which is difficult to resolve in reality.

The second reason underpinning the lack of institutional investment in the market is the lack of transactable stock. After all, there is still 2.34 million sq m of Grade A or higher quality single-ownership stock in Dubai. However, only a very small portion of this stock has been transacted, largely due to the reluctance of owners to want to sell such stock (as it is limited in supply).

One factor behind this reluctance has been the downward pressure that commercial real estate markets have faced over recent years. Since 2015, declining performance indicators have meant that valuations of these assets have decreased rapidly and therefore any transactions would likely be in the favour of the buyer. More so, from an owner's perspective, the income being generated by these assets, despite declining performance, was preferred to realizing a discounted sale and searching for or developing new assets in a relatively challenging market. In effect, there has been very little opportunity cost in not bringing quality assets to the market, which has underpinned the lack of activity. As a result, capital which had been looking for such exposure has been exported to a range of global markets.

What is driving the future demand for institutional investment in Dubai?

However, over the last year, there have been a number of changes which could be a catalyst to growth in investment volumes in Dubai's commercial real estate space going forward.

First, we have seen a number of changes to regulation both in Dubai and indeed the wider UAE. The ease of doing business, foreign ownership of business, changes to residency regulations and easing of geopolitical tensions have and will continue to change how foreign capital perceives the UAE and in turn its main business hub, Dubai.

Second, the UAE's handling of the COVID-19 pandemic has been acknowledged as amongst the most successful globally. These changes and factors have, over a relatively short period, provided a significant boost to both the UAE and Dubai. It is important to also note that this is all even before the benefits of EXPO 2020 are being materially realized.

In turn, we have also seen Dubai's real estate market enter a new market cycle, after over five years of downward pressure. Dubai's commercial market is seemingly also, at least in parts, beginning to enter a new market cycle. This shift in regulatory fundamentals and market performance may spur demand for the delivery of a new generation of quality commercial real estate. As market performance improves and demand for greater quality commercial real estate increases, where premiums can be achieved, the opportunity cost of not developing such assets increases. With valuations for existing assets likely to increase, where this will be most evident initially for Grade A and higher quality buildings, owners of such assets are much more likely to market their assets in the coming years. In addition to this, with recent changes in regulations where owners can enact higher rental increases if they have or invest in the quality of the amenities provided, as a result of this, we may also see an even greater increase level of investable stock increase going forward.

As a result of these changes and the ever-increasing focus to diversify the economy, where outbound investment from public sources has been largely diverted, we expect that Dubai's commercial investment market will see a marked increase in activity level. Whilst, initially, activity will originate from local and regional capital, as transparency improves and investment sentiment improves, we are likely to see greater levels of interest and participation from international capital.

Dubai's model of "Build it and they will come" has certainly served it well and by no means is it over, it is simply impossible to keep the earlier level of pace. This forecast of increased and increasing activity levels will be part of the next phase in keeping capital in Dubai for the long run.

The impact of iconic towers and property investment activity

The role of tall towers, urban regeneration and property investment is well documented globally. Chapter 1 examined the development philosophy behind tall buildings, and there would be a clear line of argument to also suggest the impact of tall buildings on property investing, particularly those of a highly iconic nature. In Dubai, the Burj Khalifa has undoubtedly positioned Dubai as a global hotspot and has positioned Downtown Dubai as an epicentre of global tourism. Box 2.3 refers to the fact that the tower is a part of the large downtown masterplan, but as with most tall towers it is symbolic for Dubai and the master community it sits within, adding to the experience of visiting and living in Downtown Dubai. Dubai Creek Harbour, one of Dubai's other main master-planned communities will be home to a new taller tower, confirmation of a new urban growth centre for Dubai. Box 2.3 elaborates on the impact a super-tall tower might have on the surrounding area

---

Box 2.3: The impact of the Burj Khalifa for property investment in Dubai

Burj Khalifa is the centrepiece of a 2 sq km, US$20 billion (AED73 billion) mixed-use development known as Downtown Dubai developed by Emaar. The rise of the Burj Khalifa as the world's tallest tower in 2010 was symbolic of the fact that Dubai was open for business and it has become synonymous with Dubai's growth in tourism. This is somewhat supported by the neighbouring Dubai Mall, the world's largest retail space. The US$1.5 billion construction of the world's tallest tower began in 2004 and was completed in 2010. But how has it performed as an investment over the last decade since its inception?

What impact has it had on prices in the surrounding area (Downtown Dubai)?

From 2002-2008, availability of finance and rapidly rising rental yields kickstarted the first wave of property price speculation in Dubai. Prices in the Burj Khalifa between 2004 and 2008 rose from AED1,200 psf to AED4,500 psf, an annual price appreciation of almost 40%. Rental yields had gone from 10%–12% per annum to c. 7%–8%. This drew capital appreciation. In 2009, the Global Financial Crisis halted the euphoria and prices plummeted to AED2,000 psf falling further to AED1,400 psf in 2010. Over the next ten years (2011–2021), the prices in Burj Khalifa never recovered back to their peak yet they steadily settled around the AED2,000 psf mark. Between 2004 and 2021, prices have risen by 46.25%, an average of 2.3% annually. As the Dubai market has matured, prices have stabilized somewhat, but another boom-bust cycle followed in the 2010–2020 period, albeit to a less volatile extent (see Figure 2.2). Average sale prices between the Burj and Downtown have also converged (Figure 2.3).

Table 2.4 does show that the Burj and Downtown Dubai do seem to be trading at a significant premium when compared to the Dubai city average, ranging from 75% to 125%. Dubai's high-rise construction has

clearly served well as a symbolic economic function in the city's ongoing aspiration to attract both human and financial capital.

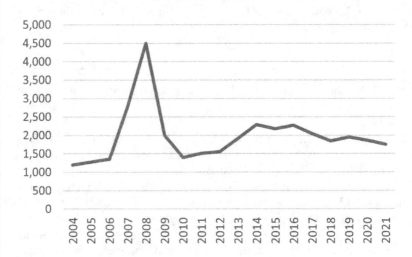

Figure 2.2 Burj Dubai Prices per sq ft (2004–2021).
Source: supplied by Andrew Baum, cited in Baum (2022)

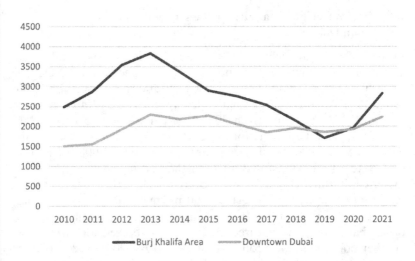

Figure 2.3 Average sale prices between Downtown Dubai and Burj Khalifa, price per sq ft.
Source: rounded data from Data Finder (2022).

Table 2.4 Average (median) sales prices per sq ft: Burj Khalifa vs Downtown Dubai

Type		2010	2011	2012	2013	2014	2015	2016	2017	2018	2019	2020	2021
Apartments	Burj Khalifa	2,395	2,490	2,870	3,535	3,830	3,375	2,900	2,750	2,540	2,145	1,715	1,955
	% change		4.0	15.3	23.1	8.2	−11.8	−14.2	−4.9	−7.8	−15.7	−20.0	14.0
	Downtown Dubai	1,400	1,510	1,555	1,915	2,290	2,180	2,270	2,050	1,850	1,950	1,860	1,755
	% change		7.9	3.0	23.2	19.6	−4.8	4.1	−9.7	−9.8	5.4	−4.6	−5.7
	Dubai (All)			890	950	1,040	1,000	1,000	1,040	970	856	833	966
	Downtown Premium			75%	102%	120%	118%	127%	97%	91%	128%	123%	82%

Source: Data Finder (rounded), Michael Waters.

The next section will examine the how of investing in Dubai's real estate sector from a broad approach. It also highlights the merits of investing and what form investments in Dubai's property sector have typically been made.

How can property investors invest into Dubai's property market?

Property investments can be split into two broad categories:

* *Direct investment* which is the ownership of the physical asset, for both owner-occupation (end-user) and as an investment (to let).
* *Indirect investment* is a paper asset backed by property. Property company shares and real estate investment trusts (REITs) are examples of indirect vehicles. It has been popular historically for investors to buy shares in large property development companies such as Emaar, UPP, Damac and Deyaar. Some listed property stocks are limited to GCC investors only. Emirates REIT was also introduced in 2010 and provides investors exposure to a diversified portfolio of commercial assets (such as schools; industrial; retail and offices). Retail investors can also participate in crowdfunding platforms, such as Stake (www.getstake.com) or Smart Crowd and get exposure to the Dubai residential market with as little as AED5,000. Investors are collective parts of a special purchase vehicle (registered) and make returns based on the rent received from the selected property. More recently in 2022, Dubai offered fractional ownership for up to four individuals to purchase a property in Dubai. This can apply to new and secondary property purchases in Dubai (subject to the approval of Real Estate Regulatory Agency (RERA)/Dubai Land Department (DLD)). According to secondary sources, fractional title deed owners can sell their stakes and transfer the deed to another buyer (most likely subject to an independent property valuation at the time of transfer). Dubai now offers investors in 2022 a much broader base from which to purchase real estate when compared to earlier years. Table 2.5 shows a summary of the main advantages and disadvantages of direct and indirect property ownership.

How are property purchases financed in Dubai?

While globally, most purchases in real estate are financed (mortgage-backed), the majority of Dubai transactions have been cash-based transactions. Table 2.6 shows a record of annual property transactions since 2015. During this period, cash purchases appeared to be a consistently significant proportion of property transactions, ranging from 59% to 71% of registered DLD transactions.

These statistics may support the popular belief that Dubai notably attracts a greater proportion of HNWI foreign investors and may indicate

Table 2.5 Summary of direct and indirect property investment in Dubai

Direct property investment	
Advantages	**Disadvantages**
Physical, tangible asset	
Fixed increase/index-linked (rents)	Imperfect information (decision-making)
Capital appreciation	High transaction costs
Contractual income (lease)	Management costs
Provides diversification (in a financial portfolio)	Diversification within direct property is hard due to large capital required
Lower volatility (compared to other assets)	Depreciation
Indirect property investment	
Advantages	**Disadvantages**
Smaller financial investments ("fractions")	Lack of control of the investment (stock selection)
Potential for greater liquidity	
No direct property management required by investor (via fund, specialist)	Entry and exit fees plus annual management fees (as a %)
Provides opportunity for greater diversification across multiple real estate asset classes	Potential for lock-ins (fixed holding periods)
	Uncertain/un-tested resale potential
	Property within the vehicles still depreciates

Table 2.6 Annual property transactions

Year	First sale	Secondary sales	Total	Mortgage	% mortgage	% cash
2015	28,907	18,726	47,633	14,059	29.5	70.5
2016	23,963	18,040	42,003	14,397	34.3	65.7
2017	30,969	17,207	48,176	15,538	32.3	67.7
2018	21,015	12,864	33,879	13,960	41.2	58.8
2019	27,972	12,041	40,013	12,492	31.2	68.8
2020	10,096	25,245	35,341	12,957	36.7	63.3
2021	36,491	24,434	60,925	19,507	32.0	68.0

Source: data sourced from DLD (2022).

why cash purchases have consistently remained the majority, despite a low-interest rate mortgage environment. Dubai has emerged as one of the most popular cities in the world for ultra-wealthy families, as explored in Chapter 1, a likely beneficiary of the global growth of UHNWIs increasing in nearby India and China.

Some analysts also view the data in Table 2.6 as a proxy for owner-occupier trends in Dubai (% of end-users matching the % of mortgage buyers). Though this might not be wholly accurate, it may provide a suitable indication (given the lack of collected data on property ownership rates). If assumed to be the case, then owner-occupier rates in Dubai are significantly lower than in other global markets. For instance, home ownership rates in the UK, Australia, the US and the Eurozone are c. 65%; notably, Germany, Switzerland and Hong Kong have lower rates at around 50%. Several new long-term population demand drivers have been introduced to attract and retain global talent, which may see home ownership rates increase. These include:

- Ten-year ("golden") visa for skilled professionals and business entrepreneurs
- Retirees (over 55 years old) can obtain a five-year residence visa (depending upon income and asset values)
- 100% corporate ownership laws expected to drive the creation of new business set-up: employment and economic growth

Furthermore, the recent change in the decriminalization of bounced cheques/defaulting may encourage or support more mortgage buyers entering the market. From January 2022, a criminal liability on bounced cheques/default is no longer applicable if the cause behind dishonouring the cheque is insufficient funds (unless it is done with intention). Therefore, debt taken on a property now starts to be considered with more willingness and may bring new entrants to the market. This coupled with the lower 20% loan deposit requirements and lower interest rates are likely to support transactional activity. We might also expect to see a greater proportion of investors looking to purchase property in Dubai using a mortgage. Table 2.10 shows why buying with a mortgage can also make more financial sense.

How is commercial real estate priced?

Pricing of real estate is always about capturing people's/firm's willingness to pay based on risk and return opportunity. It is of course a subjective assessment of a future cash flow either explicitly mapped out from a lease or assumed to fit a particular holding period (typically via a discounted cash flow (DCF) analysis). Subjectivity and future forecasting mean that pricing will rarely be consistent between parties and the different investors will be willing to pay either more or less than a competing bidder. Of course, in bidding, the property gets sold to the highest bidder regardless of the specific mechanisms used to assess the risk and opportunity of a real estate acquisition. Key components of a DCF analysis would include:

- What is an appropriate discount rate to assume when purchasing property?
- How accurate are your cash flow assumptions?

- If you assume a ten-year holding period, how do you account for your cash flow assumptions when leases typically reach 1–3 years?
- How is rental growth stated in the lease and/or your assumptions?
- What assumptions are you making about your property's terminal/exit value?

When appraising a property acquisition from such a list of questions, the two key areas of subjective difference would be first, the applied discount rate and second the exit yield. There is no scientific equation that fits perfectly to universally deciding on the "right" discount rate or exit yield. Instead, it is based on an evaluation of risk. If I am investing in low-risk assets/investments that yield 5%, I would be seeking a higher discount rate on a riskier asset to compensate me for the additional risk and uncertainty I might be taking on when I acquire the property. Do I add 1% or 4% to that discount rate? A competent professional would look to source comparative transactions to see market evidence as to what similar assets are being traded in the market currently. It does not mean however that simply observing a transaction which has an initial yield of 7% that I should also acquire on this basis if I felt the risks of acquisition of the specific property did not warrant a 7% discount rate (or desired/target return). Yet, it does tell me how my competing bidders are thinking currently in the market. It is seldom useful to run an appraisal applying a 12% discount rate, when the rest of the market is paying 7%. I will simply be outbid as my higher discount rate would translate into a lower net present value and a lower offer price for the asset. Take the assumption that I have the capacity to purchase a building that has a net annual income of AED1,000,000. Table 2.7 illustrates how the discount rate can significantly impact the specific price I would be willing to pay for the asset. Rental growth is fixed for simplicity in this example. The property is freehold so assumes a terminal value (sale value) at the end of Year 5, which would be based on the annual rent/assumed exit yield. Again for simplicity, I will assume the rent stays fixed and the exit yield is assumed to be 5% (or a cap rate multiplier of 20, expressed from 100/5). Based on the assumed rent, this provides a future sale value of AED20,000,000 (AED1,000,000 × 20).

This simple analysis shows that the subject property will have a range of values significantly different based on the assumed discount rate by the potential buyer, influencing the price/value of the property by as much as 30% in this case example. The discount rate assumed will reflect the individual's bid and assessment of risk. In a commercial property this would surround the quality of the lease terms (rent; rent escalations; cost responsibilities); the quality of the tenant who is paying rent (default risk) and the grade of the asset itself. Grade A assets (prime location, modern, single owned, long-lease, top quality tenant/s) will carry a lower risk than if the property was Grade B or C (older, some vacancy risk; poorer quality location and tenant mix; deferred maintenance; obsolescence). A prospective purchaser would

Table 2.7 Present value ("Offer bid") of a property acquisition based on different discount rates

Year	0	1	2	3	4	5	NPV
Net annual income	1,000,000	1,000,000	1,000,000	1,000,000	1,000,000		
Sale Value						20,000,000	
Discount rate 5%	1,000,000	952,381	907,029	863,838	822,702	15,670,523	20,216,474
Discount rate 8%	1,000,000	925,926	857,339	793,832	735,030	13,611,664	17,923,791
Discount rate 12%	1,000,000	892,857	797,194	711,780	635,518	11,348,537	15,385,886

NPV: Net present value

Table 2.8 Common considerations in risk-adjusting a discount rate or yield

Risk variable	Increase discount rate/yield	Decrease discount rate/yield
Potential for future growth	Low	High
Economic conditions	Poor	Strong
Market uncertainty	High	Low
Location	Poor	Strong
Level of competition (supply)	High	Low
Financial uncertainty	High	Low
Legal uncertainty (lease renewals)	High	Low
Tenant quality	Poor	Strong
Physical obsolescence	High	Low
Occupational uncertainty	High	Low
Leasing uncertainty (lease length, voids)	High	Low
Data/valuation uncertainty (lack of comparables)	High	Low

Source: Author's own

make some positive (lower) or negative (higher) discount rate adjustments in their appraisal to factor in these specific risks. Table 2.8 highlights a wide range of common considerations related to adjusting the discount rate for an investor.

I will now provide an example of how an individual (expat) may appraise a residential property purchase and conclude on the things to consider when buying in Dubai. Further details on these topics are also discussed in Chapters 3–5 detailing the residential sector and commercial sector more specifically.

Buying residential property in Dubai: An example appraisal

As with most investments made, an appraisal is often necessary to make sense of the purchase or investment decision. The next section provides an overview of how an investor may choose to appraise a direct property purchase in Dubai. It takes on the case of a residential property purchase. When making an investment into real estate, an investor is having consideration to two main components:

1   The cash flow (income return), and future returns are not guaranteed (unless they are explicitly mapped out in a long lease) and so an investor will also evaluate the income security (risk) expressed through a yield (or cap rate).
2   The time value of money (relative risk/opportunity cost). Modern-day investing and investment management is about enhancing the return while managing risk. Investors also need to allocate funds to assets and they have finite resources so decisions are based on a number of risk considerations (liquidity; marketability and management costs). The ease of management also needs consideration above and beyond pure financial returns.

Case study: A residential property investment appraisal technique

This case study considers the financial appraisal of residential property investment in Dubai, taking the view of an individual investor using a DCF appraisal technique. This analysis considers a direct residential property investment.

Scenario

An expat is considering purchasing an investment property with an asking price of AED1,550,000 in a popular freehold master-plan community (comprising of 1,300 sq ft). The investor plans to let the property for five years before selling it (I would often advocate a ten-year hold on any property appraisal, but this example is only for demonstration purposes). Currently, this property will lease for AED85,000 per annum and service fees/management are AED17.50 per sq ft. The expat is looking to take a mortgage and contribute a 20% deposit (the current minimum deposit required in Dubai). The mortgage rate can be fixed for five years at 3% per annum. The RERA rental index indicates that rents for this type of property currently range from AED100,000 to 120,000 per annum. Therefore, the rent can be increased by AED8,500 at Year 2; AED4,675 at Year 3; AED4,908 atYear 4 (until reaching the current market rent stabilization). The expat is expecting a target rate of return based on 5% and so assumes this as an appropriate discount rate.

Key appraisal assumptions

Initial purchase price: AED1,550,000, equity deposit of 20% (AED310,000)
 Upfront purchase costs: AED118,370. When buying a residential property in Dubai, there is a range of purchase costs that prospective buyers should be mindful of in their buying decision. Below is a summary of these costs and a breakdown of how it is calculated:

Land Department fee (4% + AED540)	*AED62,540*
Title registration fee (AED2,000 purchase of AED500,000 or less; AED4,000 above AED500,000)	AED4,000
Mortgage registration fee (0.25% of loan + AED10)	AED3,110
Agent fee (2% purchase price) + 5% VAT	AED32,550
Bank arrangement fee (c. 1% loan amount) + 5% VAT	AED13,020
Valuation fee + 5% VAT	AED3,150
Total purchase cost	AED118,370
As a % of the purchase price	7.64%

- Annual mortgage cost: AED72,781 (including mandatory life and property insurance premiums). Mortgage assumptions based on an annual mortgage payment calculation of AED71,211, plus AED1,570 per annum for mandatory life and property insurance. The formula for calculating annual mortgage payments is shown in Chapter 3 ("Cash vs mortgage purchase analysis")
- Annual service charge: AED22,750
- Annual rent: starting at AED85,000 with escalation as per the rental index
- Capital value after a five-year holding period: AED1,980,000 (based on an assumed 5% pa growth rate). In reality, Dubai house prices have grown by 2.1% pa (10-year) and 3.7% pa (20-year) and I would recommend the use of growth rates in this range (see Chapter 4)
- Mortgage outstanding at the end of year 5 is AED1,060,269. Therefore, the buyer is expected to receive a capital balance of AED919,731 at the resale (based on future sale price minus outstanding debt, 1,980,000-1,060,269).

The DCF analysis begins by examining the proposition to buy with a cash purchase. Table 2.9 presents a suitable analytical DCF framework for doing so.

Whilst many investors use leverage to raise the capital to purchase a property, it can also be used to improve their return on equity within a property acquisition (also allowing for diversification of funds within a real estate portfolio rather than locking it up in a single property). Table 2.10 takes the analysis of the same acquisition this time with a 20% equity down payment/80% finance. This currently represents the lowest amount of deposit required in Dubai.

Table 2.9 Residential appraisal with cash purchase

Year	1	2	3	4	5	6
Equity contribution	−1,550,000					
Purchase fees	−99,090*					
Mortgage payment						
Service fee	−22,750	−22,750	−22,750	−22,750	−22,750	
Income	85,000	93,500	98,175	103,083	103,083	1,980,000
Disposal costs						−39,600
Net cash flow	−1,586,840	70,750	75,425	80,333	80,333	1,940,400
PV @5%	1	0.9524	0.9070	0.8638	0.8227	0.7835
Discounted c/flow	−1,586,840	67,381	68,413	69,395	66,090	1,520,354
NPV	**204,793**					
IRR	**7.80%**					

* purchase costs when buying with cash are reduced to 99,090 as there is no mortgage registration fee; no bank arrangement fee and no valuation fee

Table 2.10 Residential appraisal with 20% LTV assumed

Year	1	2	3	4	5	6
Equity contribution	−310,000					
Purchase fees	−118,370					
Mortgage payment	−72,781	−72,781	−72,781	−72,781	−72,781	
Service fee	−22,750	−22,750	−22,750	−22,750	−22,750	
Income	85,000	93,500	98,175	103,083	103,083	919.731
Disposal costs						−39,600
Net cash flow	−438,901	−2,031	2,644	7,552	7,552	880,131
PV @5%	1	0.9524	0.9070	0.8638	0.8227	0.7835
Discounted c/flow	−438,901	−1,934	2,398	6,523	6,213	689,583
NPV	**263,882**					
IRR	**15.43%**					

Based on this initial analysis, if the expat was to offer to buy the property at the asking price of AED1.55 million and rent it for a holding period of five years, he would get an annualized return of 7.80% per annum if purchasing in cash and 15.43% when buying with a mortgage. Leverage as with any real estate investment enhances the return on equity contributed. The annualized unleveraged return in this example shows a stabilized rental yield of 6.7% (gross) and 5.2% (net). This is attractive when compared to other global markets that might see rental yields in the region of 2%–4%. In Dubai, purchase costs are also reasonable, in this example totalling 7.6% of the buyers cost (mortgage) and 6.4% (cash). Nowadays, some banks are willing to wrap these fees up in the mortgage so a buyer is only left with raising the 20% deposit. If this was assumed in the analysis the IRR would be higher. When undertaking an appraisal, it would be expected to run a series of scenario testing. This could include asking the following:

- How reasonable is the AED1.55 million asking price if the investor now seeks a 7% target rate of return, what would the maximum purchase price be?
- Over a longer holding period, it would be reasonable to assume an increase in service/management fees or mortgage rates and these should be stress-tested in any appraisal.

I have highlighted a range of assumptions that would form part of a prudent buyer's analysis and run some simple sensitivity testing to evaluate whether the purchase is still reasonable. I have focused the analysis on changes to the following:

- **A change to the interest rate.** Currently, a buyer might expect the cost of their mortgage to increase beyond the five-year fixed rate period. A mortgage buyer should be mindful to review the bank's fixed rate margin once the mortgage moves to a variable rate. Typically, in the final mortgage offer, a bank will present a fixed rate + three-month (3M) Emirates Interbank Offered Rate (EIBOR). This will be used to calculate the revised interest rate when the fixed-rate period expires. Let us assume this is 2% + 3M EIBOR. Based on this, we can carry out a simple analysis:

Fixed rate (for the first five years)	3% (current rate fixed until 2027)
Bank's variable rate fixed margin	2%
Current 3M EIBOR	0.67% (as of 7th March 2022)
Variable rate (assuming todays rates)	2.67%
Revised 2027 rate (assuming 1.5% base rate increase)	2.67 +1.5 = 4.17%

EIBORs are published daily on https://www.centralbank.ae/en/services/eibor-prices

In this analysis, it might be prudent for the buyer to stress test their mortgage costs after five years to assume a 4.5% interest rate on their mortgage. If after five years the mortgage balance will be AED1,060,269, then the assumed annual mortgage costs when applying the new variable rate of 4.5% would be AED81,509/annum, an 12% increase from the fixed rate costs. Some banks/analysts would take it further and stress tests the mortgage by adding 5% to the current rate to check the mortgage affordability. If we assumed, mortgage rates went to 8%, then the annual mortgage cost would be AED107,991/annum (a 48% increase). Whilst this might be unlikely, it would be prudent to see what the mortgage running costs would be if the market saw a large interest rate hike and for the buyer to evaluate how these costs could be covered if that was to occur.

- **Service fees/management costs.** It might be reasonable for the purchaser to factor in an increase in the service fees/management costs after the first five years. Real estates are durable assets, yet buildings do depreciate over time and it would be prudent to factor in a 10% increase in service fee rates to factor in any longer-term maintenance costs that may occur beyond 5-, 10- or 15-year hold periods.
- **Future rental growth and sale price.** This figure is assumed off a historical growth rate or in the case of a commercial building would be calculated off future rent × future cap rate multiplier (sometimes termed as an exit yield or a terminal yield). This figure needs stress testing as it is based on a foreseeable future value and is not a guaranteed sale price.

These parameters would form part of a prudent buyer analysis for either a residential or commercial asset in Dubai. Further examples are presented in later chapters.

Conclusion

This chapter has aimed to provide an overview of property market activity in Dubai. Over recent years, there has been a range of new initiatives introduced that address the previous criticisms of investment into Dubai's real estate market. These have included: rental caps; lease registration and generally better levels of statutory and consumer protection, especially those relating to off-plan sales (with the use and implementation of Escrow accounts). Alongside the legal aspects, the Dubai property market also witnessed an exceptional amount of investment demand from 2002 to 2008 as a result of its open-border policy to foreign investment in key designated areas across the Emirates, known as "free zones". Residential investment has been the dominant recipient of foreign capital, attracting funds from GCC and international HNWIs. However, the range of investment products and locations is still somewhat limited for investment in Dubai's commercial stock, largely as institutional Grade A office supply is somewhat limited. Therefore, much of the foreign investment demand has remained for the residential sector.

In 2021/2022, Dubai is still somewhat of an attractive proposition, providing investors with high annual returns and scope for capital appreciation. Further global commentators are starting to see "value" in Dubai, evidenced in the UBS report stating Dubai as an "underpriced" market, suggesting an imbalance between market values and rational pricing theory. Yet pricing has historically been a function of demand and supply, rather imperfect given the opaque market data available (see Chapter 8). If markets are opaque, as historically in Dubai, there have been opportunities for investors to make excess returns over the risk-free rate for investors choosing to invest in Dubai's property market. One viewpoint is that as global property markets converge the opportunity for excess returns diminishes (a reduction in geographical arbitrage) and we are likely to see this play out in the Dubai market too. On the flip side with better information and improved market transparency, a path towards more institutional investment is likely to be more apparent. As the city and its property market evolve, it is likely to carry less risk, first demonstrated through a reduction in financing rates and subsequently in property yields. For those who entered the Dubai market in the previous cycles, this is likely to provide a strong capital appreciation and upward prices on a more sustained and long-term perspective (rather than one on speculation over a short number of years). Furthermore, global investors are now seeking diversification benefits away from traditional core markets. While international real estate investment offers attractive return opportunities for investors, there is no "free lunch" and the risks of international investments should be made apparent.

Chapter 4 presents in more detail the issues that arise and are likely to need managing when investing in Dubai as part of a global property portfolio. The next chapter moves into a broad analysis of Dubai's residential market.

References

Baum, A. (2022) *Real Estate Investment: A Strategic Approach* (4th ed.), London: Routledge.

Data Finder (2022) https://www.datafinder.ae/.

DLD (2022) Yearwise Breakdown of Transactions. https://dubailand.gov.ae/en/eservices/real-estate-transaction/#/ (Accessed March 2022).

Dubai Municipality (2021) 'Dubai Building Code: 2021 Edition, Dubai Government, https://www.dm.gov.ae/municipality-business/planning-and-construction/dubai-building-code-2/

Real Capital Analytics (2021) https://www.msci.com/our-solutions/real-assets/real-capital-analytics.

REIDIN https://reidin.com/.

Savills (2021) The Total Value of Global Real Estate. https://www.savills.com/impacts/market-trends/the-total-value-of-global-real-estate.html.

Part B

Analyzing the Dubai property markets

3   Residential

This chapter has been written to provide some key insights into the Dubai residential sector. It is perhaps the most discussed segment of the market and has received the most global attention over time since the inception of foreign ownership laws in 2002. The laws and legislations are well established. For example, Real Estate Law No. 7 of 2006: Land Registration Law in the Emirate outlines that a UAE citizen and GCC citizen can buy real estate anywhere in Dubai, yet foreign nationals can only buy in designated "freehold" areas via either freehold or leasehold ownership structures. Many people still have reservations about buying in Dubai, even those that have been living in Dubai for 10–15 years and beyond. I often meet a lot of new expats in Dubai who perhaps do not appreciate the historical context of Dubai's residential sector. This chapter therefore begins by presenting a time series of residential price movements from 2002 to 2021. The property cycle of Dubai's residential sector is explored in detail and examines the key forces that have driven the rise and fall of residential property prices over time. After examining the historical context, it then examines the property buying processes unique to Dubai and establishes a summary of the current agency laws and practices for buying residential property in Dubai (see Chapter 6). Whether you are looking to buy a two-bed apartment in Dubai Marina or Downtown Burj Khalifa or you are considering that large, detached villa, buying property in Dubai, like elsewhere, is often one of the largest personal investments people make, so how can you get this right? This chapter examines some simple strategies for you to start your own residential portfolio in Dubai. The content examines leverage from a historical development perspective, both in terms of regulation and investor returns. I highlight some of the common requirements relating to the use of leverage when considering to invest in Dubai real estate. The latter sections of this chapter illustrate where is there scope to maximize your investment returns via leverage; what to look out for when securing a mortgage; as well as a suggested mortgage repayment strategy that is based off rental cover.

DOI: 10.1201/9781003186908-5

First, I would like to drawdown on some key statistics that are in support of residential property investment in Dubai:

- Dubai's population has grown by a Compound Annual Growth Rate (CAGR) of 6.5% over the last five years. It is expected to reach 5.8 million by 2040
- Over 90% of Dubai's population are expatriates. Dubai's large young demographic profile and growing expatriate community has intensified the demand for rental property. Approximately 80% of Dubai's expatriate resident population is part of the private rented sector (with an assumed owner-occupation rate of approximately 20%).
- Dubai's 2040 masterplan has been set to enhance the city's liveability by developing and investing further in transportation, education and healthcare sectors.
- In addition to the above, government regulatory reforms are expected to further strengthen the economy and market confidence, hence increasing investment into the sector.
- According to Dubai Statistics, at the end of 2021, there were 755,074 residential units in Dubai's urban areas (72% apartments and 18% villas). In 2021, +20,000 villas were added versus a 5-year average of +6,750.
- The 5-year average residential build rate (2015-2020) has been +35,210 units/annum. It is anticipated Dubai will need an average of +27,500 units/annum to accommodate the 2040 population forecast of 5.8m people.

Despite these points, the history of investing in Dubai's property market has been a bumpy ride. While a relatively new market, there have been three separate property cycles in Dubai, with the last two cycles spanning approximately nine years (see Table 3.2). The next section examines in more detail an overview of one of the world's most fascinating property cycles.

Introduction: The Dubai residential property cycle

The globalization of Dubai's residential market was established in 2002, with the introduction of foreign ownership laws (as previously outlined in previous chapters). Since then, the Dubai residential cycle has experienced two peaks. As suggested in Chapter 1, with real estate a significant GDP contributor then new development activity is correlated to GDP growth, as with other global markets. Table 3.1 begins by outlining the historical time series of Dubai's residential cycle.

Figure 3.1 illustrates the Dubai residential cycle graphically. From these data, which begins in 2008, we can observe that there have been two residential cycles (one peaking in mid-2008; the other peaking in mid-2014). We are now entering a third cycle, coming off price declines that are showing signs of strong recovery since late 2021.

Table 3.1 Overview of the Dubai property cycle

Date	Stage	% change in sale price	Market conditions
January 06/August 08	Rapid growth	+92	Excess liquidity and limited supply
August 08/January 11	Correction	−36	GFC and local debt crises
January 11/May 12	Stabilization	+10	Gradual return of confidence
November 12/August 14	Rapid growth	+61	Market growth/euphoria following the Expo 2020 announcement
August–December 2014	Peaking out	+10	Continuation of growth trend but slower place
January 2015–December 2019	Soft landing/ gradual correction	−25%–30% declines over the period	Oversupply commentary based from an overconfidence in development of new affordable housing and financial payment plans from developers
February 2020–September 2020	Further correction due to COVID	−10%–15% further declines	The global impact of COVID-19 led to further downward pressure on values
November 2020–January 2021 +	Upward price increases	Selective recovery (c. 5%–10%)	Pent up demand and high transactional activity saw a slight increase in property prices
January 2021–December 2021	Further price recovery	Residential prices increased by an average of 9.2%, the quickest pace of growth since January 2015	Race for space drove up significant price growth in villa communities in particular (21.2%)

Source: Waters (2019). Author has added additional commentary to account for more recent movement in Dubai property since August 2014.

Amid the Global Financial Crisis (GFC), average Dubai residential property prices fell to c. AED785 psf. House price declines were initially recorded in Q3/Q4 of 2008, with an average price decline of 8%. Despite this initial decline, property prices stood at 60% above the previous 12 months. Q1 2009 measured deeper price declines of 34% (bringing them back to Q2 2007 levels (c. AED1,042 psf)). Colliers (2009) reported a blended average of AED1,037 psf in Q12009, down from AED1,770 psf in Q42008 and villas and townhouses also saw a 40% drop in value. The decline in GFC property prices was attributed largely to the sharp reduction in the availability of finance. Two of Dubai's largest mortgage providers, Amlak and Tamweel had withdrawn from

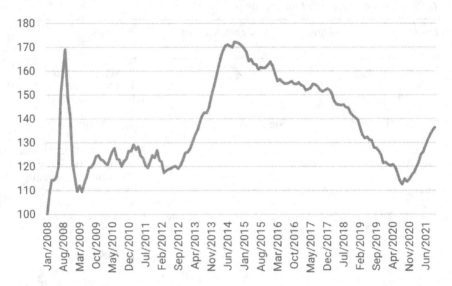

Figure 3.1 Dubai's dynamic house price index (2008–2021).
Source: Property Monitor (2022), data supplied by Zhann Jochinke

the market. Borrowers had to now raise 25%–50% deposits to buy, up from the 5% to 10% previously required. In 2011, LTV deposit requirements had been tightened to 30%–35%, significantly higher than many other global markets.

By mid-late 2011, there were anecdotal signs of stability and small price growth, however, during this period a variety of externalities caused friction to a quick-paced recovery most notably the Arab Spring. Apparent decreasing vacancy rates during the recovery phase (2011–2012), saw supply in the market reduced slowly as the new development had ceased with low-moderate employment growth. An expansion period between 2013 and 2014 saw transaction prices begin to rise as the impact of the GFC had subsided globally and the local economy saw strong signs of recovery. Local bank lending began to normalize, and buyers were once again permitted to 20% LTV on residential mortgages. At the same time, lending rates had dropped 200 basis points compared to 2009, and alongside lower prices, the Dubai residential market looked attractively priced. Residential property prices for Dubai averaged c. AED1,250 psf, in par with the previous peak of 2008. As prices peaked towards the latter part of 2014, government measures had been put in place to cool once again a speculative run up in property prices. These measures included a doubling of property transfer fees from 2% to 4% and an increase in LTV (25% on first purchase; and 60% on subsequent property purchases). At this time, oil prices had dropped significantly shaking the foundations of the local oil economies. In parallel to this, developers had been launching an aggressive development campaign during 2013–2014 built around affordability in the new suburban residential communities of

Table 3.2 Summary of the three main residential cycles in Dubai

	Start (Year)	AED psf	Cycle peak	% increase	Duration
Cycle 1	2002	500	1,309 (2008)	162	8
Cycle 2	2011	832	1,487	79	9.8
Cycle 3	2020	998			

Source: Knight Frank (2022).

Dubai. Significant oversupply of residential units entered the market periodically during 2015–2019, which saw a gradual and extended decline in property prices over that 5–6-year timeframe. Recent market data suggest that more stable market conditions now exist in late 2020/2021. At the macroeconomic level, this has been attributed to both better real estate regulation and funding restrictions across the industry. Liquidity has fed into the local real estate market in a similar way it has done so in other global markets, with LTV reducing back down to 20%. Dubai is also experiencing historically the lowest level of mortgage interest rates seen since 2002 (it was not uncommon to see mortgage rates at 8%–9% in 2008–2010; now we are seeing rates as low as 2.5% in 2021). This has stimulated demand from the expatriates who once again are now seeing Dubai priced attractively and as a longer-term residence, with buying seeming a more preferred option to long-term renting. Current residential prices in 2021 in Dubai are on average c. AED900 psf. According to Knight Frank (2022), Dubai's residential market recorded over 52,000 apartment and villa transactions, totalling AED114.2 billion, more than the total for 2019 and 2020 combined. Table 3.2 summarizes the three main cycles seen in Dubai's residential sector.

How leverage impacted the Dubai residential market pricing?

One of the key price growth drivers for any real estate market is the availability of finance. When interest rates are low and LTV ratios are lowered, we tend to see price growth. This is based on the fact that the financial requirements to enter the market are lowered and the running of a mortgage becomes more affordable, at times often significantly lower than the monthly rental payments for a similar property. Yet in Dubai, we have not historically seen a large uptake in residential mortgages. Cash purchases still represent the majority of the share when examining annual property transactions. In Chapter 2, we saw this breakdown for all transactions. Table 3.3 examines these trends specifically for residential transactions.

More than 90% of Dubai's population are expatriates and their participation in the residential market has the potential to be a key determinant of housing demand in the Emirate. Data on the proportion of expatriate resident purchasers are unavailable, however, consultants have often used mortgage transactions as a proxy for evaluating these home ownership rates (Deloitte, 2021). Table 3.3 shows that on average over the last five years

Table 3.3 Dubai residential sales (2016–2020) – value, number and % mortgage

Year	Total value of transactions (AED billion)	% mortgage	Total number transactions of units	% mortgage[a]
2016	57	40.5	37,282	21.5
2017	69	38.0	45,110	19.1
2018	50.8	43.5	34,067	21.5
2019	64.3	41.8	37,027	18.1
2020	37.1 (YTD Sept)	42.0	31,158	17.1

Source: REIDIN cited in Deloitte (2021); DSC (2021).
a   Proxy for % of owner occupation in Dubai.

mortgages have accounted for approximately 20% of annual residential transactions. If we were to take this as a proxy for home ownership rather than investment purchases that historically are made in cash, we see that the rate of home ownership in Dubai is low when compared to other global locations. An independent survey carried out in Dubai also supported this level of home ownership whereby 21% of respondents were homeowners. This compares to 68% in the UK and US; and a staggering 88% in Singapore and China. This data suggests that a greater opportunity exists to attract more buyers from the end-user/resident population demographic of Dubai.

In more developed markets, demographics have a large role to play as an ageing population explains some of the higher proportion of owner-occupier trends versus that seen in Dubai, which is typically a much younger age demographic. Social and economic factors also perhaps explain why millennials and younger generations are less willing to take on the financial obligations that come with home ownership. Flexibility through renting is often seen as the preferred option. There is also a significant lack of support for home ownership as a political agenda in Dubai. Greater emphasis has been placed on attracting inward investment from HNWIs. That is not to say Dubai does not have a supportive home-ownership environment, it is just that for the larger proportion of the local population (expats), this is privately-funded through corporate employment packages rather than centrally governed as we see in other jurisdictions. As Dubai is a non-welfare state (for expats), it would be reasonable not to assume the same level of government support for home ownership. A national housing programme is available to UAE nationals. Another factor to explain differential changes in home-ownership propensities could be the changes in the relative cost of owning and renting overtime, that would see them paying a lower amount in a mortgage (plus holding costs) versus renting. Households might exhibit a higher propensity to spend more on housing (and buy) as their income increases. Other jurisdictions have seen a tax advantage of home ownership that separate the cost of owning and renting further. For instance, the UK had a period of mortgage interest relief which meant there was a tangible tax advantage (saving) compared to renting. Privatization schemes seen in the UK had a similar impact to boosting home ownership rates. Similarly in Singapore, end-users (owner occupiers) enjoy lower tax rates, perhaps encouraging home ownership. Dubai does not have such distinction in any form. The most substantial tax advantage for expat homeowners in Dubai would be the absence of capital gains tax, yet in the UK this is exempt on a primary residence, so again does little to offer supportive incentives to own vs rent. The only tangible benefit is that over time the money spent on rent is "wasted" whereas the proportion of a mortgage payment (rent) goes into paying off the principal of the loan.

One key development in the Dubai market in 2013 was the introduction of mortgage caps in residential buying. The Central Bank issued a circular to all banks and finance companies to regulate borrowing to protect both lenders and consumers by limiting the maximum LTV ratio:

- For Nationals: 80% for the first house of value less than Dhs 5 million (70% value more than Dhs 5 million), 65% for the second and subsequent houses.
- For Expats: 75% for the first house of value less than Dhs 5 million (65% value more than Dhs 5 million), 60% for the second and subsequent houses.

In 2020, the UAE Central Bank reduced the minimum deposit requirements for expatriates and UAE nationals to 20% and 15% respectively. This alongside a range of incentives offered by local banks, such as reduced or waived loan arrangement fees and lower mortgage interest rates, boosted home-buying demand. Looking ahead, an improvement in transaction volumes is predicated on demographic and economic factors alongside targeted financial plans from developers and attractive mortgage products from banks to enhance participation from both investors and end-users (Deloitte, 2021).

Box 3.1 explores how this LTV regulation may have impacted the residential property prices.

Box 3.1 The tale of two expats

The purpose of the inclusion of this case example in my book is to highlight the residential property cycle in Dubai in more detail, but also examine the impact of gearing on the volatility of residential real estate prices in two distinct time periods:

1 *Global Financial Crisis (2007–2009):* The two Expat purchases who bought in Dubai with a 10% deposit.

2   *'Expo Euphoria' (2014–2016):* The second peak in Dubai's residential property cycle, yet with much stricter financial lending
criteria, introduced in 2014 to curb property speculation, resulting in the requirement of a 25% deposit (and no further leverage
permitted on the deposit).

How did the two sets of Expat purchasers fair during these distinct
phases of Dubai's residential property pricing?
    Using Table 3.4 below answer the following questions:

a   Expat A bought a property in 2007, paying a 10% deposit for a
1,000 sq ft flat in Dubai Marina. Prices rise. The flat goes up in
value. What is his deposit worth one year later?
b   Expat B purchases a property in 2008, paying a 10% deposit for a
1,000 sq ft flat in Dubai Marina. Prices fall. The flat goes down in
value. What is his deposit worth one year later?

To illustrate the impact of price movement in each of these scenarios,
an 'equity balance' for both Expat A and Expat B can be produced,
as below:

Expat A:
    In 2007 it looks like this:
    Original purchase price: AED1,400,000
    Equity contribution: AED140,000 (the 10% deposit)
    Debt on the remaining balance: AED1,260,000
    In 2008 it looks like this:
    Revised property price: AED1,975,000
    Debt (Liability): AED1,260,000
    Outstanding 'equity' balance: AED1,975,000 – AED1,260,000 =
AED715,000– this is what his deposit of AED140,000 is worth one
year later (a return on equity of 5 times the initial deposit).

Expat B:
    In 2008 it looks like this:
    Original purchase price: AED1,975,000
    Equity Contribution: AED197,500 (the 10% deposit)
    Debt on the remaining balance: AED1,777,500
    In 2009 it looks like this:
    Revised property price: AED1,100,000
    Debt (Liability): AED1,777,500
    Outstanding 'debt' balance: AED1,100,000 – AED1,777,500 = -AED
677,500 the deposit of AED197,500 is now lost and one year later Expat
B has a debt remaining of AED677,500

Table 3.4 Dubai residential pricing (AED per sq ft) 2002–2009 (equally weighted)

Type		2002	2003	2004	2005	2006	2007	2008	2009
Apartments	Burj Dubai	–	–	1,200	1,275	1,350	2,800	4,500	2,000
	Dubai Marina	850	836	900	1,000	1,050	1,400	1,975	1,100
	Greens	500	500	725	875	950	1,250	1,700	1,000
Villas	Lakes	550	575	700	875	1,250	1,450	2,150	1,200
	Meadows	450	500	600	800	1,150	1,500	1,775	1,100
	Ranches	450	475	620	790	1,150	1,450	2,120	1,000
	Springs	420	485	500	640	1,025	1,500	1,850	1,000
Mean		536.67	561.83	749.29	893.57	1,132.14	1,621.43	2,295.71	1,200.00
Growth/loss			4.69%	33.36%	19.26%	26.70%	43.22%	41.59%	−47.73%

Source: supplied by Andrew Baum

In 2013, the UAE Central Bank raised the deposit requirements for expats buying property in Dubai to 25%. Using the data in Table 3.5 answer the following:

In 2013, Expat 1 pays a 25% deposit for a 1,000 sq ft flat in Dubai Marina. Prices rise. The flat goes up in value. What is his deposit worth one year later?

In 2015, Expat 2 pays a 25% deposit for a 1,000 sq ft flat in Dubai Marina. Prices fall. The flat goes down in value. What is his deposit worth one year later?

Again let us look at this by drawing up the same analysis for the two new Expats, Expat 1 and Expat 2.

Expat 1:
  In 2013 it looks like this:
  Original purchase price: AED1,285,000
  Equity contribution: AED321,250 (the 25% deposit)
  Debt on the remaining balance: AED963,750
  In 2014 it looks like this:
  Revised property price: AED1,585,000
  Debt (Liability): AED963,750
  Outstanding 'equity' balance: AED1,585,000– AED963,500 = AED621,500– this is what his deposit of AED321,250 is worth one year later (a return on equity of almost 2 times the initial deposit).

Expat 2:
  In 2015 it looks like this:
  Original purchase price: AED1,610,000
  Equity Contribution: AED402,500 (the 25% deposit)
  Debt on the remaining balance: AED1,207,500
  In 2016 it looks like this:
  Revised property price: AED1,530,000
  Debt (Liability): AED1,207,500
  Outstanding 'equity' balance: AED1,530,000– AED1,207,500 = AED322,500– the deposit has dropped in value by 20% to AED322,500 and there is no debt on the property.

The key observation here is how the mortgage cap has impacted price volatility. The scenarios presented in this analysis showcase how high gearing helps make high returns when capital values are rising (2007-2008) but is disastrous when values are falling (2008-2009) because investor's capital disappears (net asset/negative balance). Gearing clearly exaggerated the cycle on the upward and downward slopes. The impact of the mortgage cap has been both a reduction in price volatility and

Table 3.5 Dubai residential pricing (AED per sq ft), selected communities, 2010–2021

Type		2010	2011	2012	2013	2014	2015	2016	2017	2018	2019	2020	2021
Apartments	Burj Dubai	1,400	1,510	1,555	1,915	2,290	2,180	2,270	2,050	1,850	1,950	1,860	1,755
	Dubai Marina	880	960	1,110	1,285	1,585	1,610	1,530	1,575	1,450	1,245	1,045	1,165
	The Greens	795	710	930	1,120	1,485	1,340	1,305	1,270	1,040	900	770	835

Source: Author, data from Data Finder, rounded

lower debt exposure when prices fall. Table 3.6 also shows price movements were less volatile across the same selected communities when
compared to the earlier time period 2002-2009 (refer back to Table 3.4).

Table 3.6 Dubai residential index 2016–2021 (equally weighted)

Type		2016	2017	2018	2019	2020	2021
Apartments	Burj Dubai	2,270	2,050	1,850	1,950	1,860	1,755
	Dubai Marina	1,530	1,575	1,450	1,245	1,045	1,165
	The Greens	1,305	1,270	1,040	900	770	835
Villas	The Lakes	1,135	1,240	1,125	1,020	975	1,040
	Meadows	1,190	1,200	1,065	970	935	1,120
	Ranches	1,015	935	875	830	825	920
	Springs	795	960	895	820	735	890
Mean		1,320	1,319	1,186	1,105	1,021	1,104
Growth/loss			−0.11%	−10.88%	−6.81%	−7.63%	8.12%

Source: Author, data from Data Finder, rounded

The final section extends this discussion on mortgage finance to provide an
overview of the key choices an expat buyer can make when financing property in Dubai.

Dubai mortgages: an introduction

This section is designed to provide an overview of the mortgage product
available for a residential property purchase as well as provide a better understanding of the costs involved in borrowing for a property purchase. Historically, many property purchases in Dubai were purchased with cash, largely
this was based on the restrictive terms and rates enforced onto a property
purchaser via a mortgage. In addition, as with all debt financing, the operating cash flow is impeded by the amount of debt servicing. However, more
recently with the global fall in interest rates and increased competition and
supportive financial legislation, a higher proportion of purchases are now
being carried out with a mortgage. Interestingly, as time has progressed both
the lending rates (fixed) and the fixed-rate margin (variable) has reduced,
suggesting banks have a higher tolerance to property lending in Dubai.

Fixed- or variable-rate loans

Mortgages in Dubai can be fixed (for limited durations of up to five years)
and then revert to a variable rate. An Islamic mortgage is a "fixed payment"

for the loan duration. However, these are "rent to own" and the borrower may need to pay a balloon payment at the end of the term. It is not uncommon to see mortgages also include an interest rate floor, so regardless of where interest rates go it might state that the interest rate can go no lower than 3.5%, for instance. Borrowers should take special care in looking at this and the EIBOR margin after a fixed interest rate period expires. Typically, variable rate mortgages in Dubai are referred to as three-month (3M) EIBOR plus a fixed margin (e.g., 2%).

Conventional mortgage vs Islamic mortgage

In Dubai, mortgagor is permitted to take either a conventional mortgage or an Islamic mortgage. The main distinction between these two products relates to how the loan is structured. A conventional mortgage charges the borrower a rate of interest per year until the debt is repaid in full. Whereas, an Islamic loan does not "charge interest", instead the borrower "rents" the property from the bank for the specified loan term. Furthermore, the "rent" remains fixed for the duration of the tenure, and any outstanding balance is collected as a balloon payment at the end of the loan. An Islamic home loan differs as Sharia-compliant institutions are forbidden from charging interest and instead offer alternative financing models for home loans: Murabaha, Ijarah and Musharaka.

According to Mortgage Finder (2019), Murabaha financing is based on the bank purchasing the property on behalf of the customer and then reselling it to them for a higher price. The buyer then pays back the bank purchase price through monthly instalments. The property remains the property of the bank until the loan is cleared. Ijarah financing is a buy and lease back, more typically used when buying property off plan as no payments are made until the property is completed (Mortgage Finder, 2019). In replacement of mortgage interest in conventional mortgage, Musharaka financing charges the borrower rent based on the proportion of ownership held by both parties. Overtime the borrower buys back a greater share until owned outright (similar to a shared ownership scheme).

Lending restrictions for foreign nationals in Dubai

Since the inception of freehold property ownership in 2002 up to the GFC in 2008, banks and financial institutions had tended to offer up to 90% LTV. Since the financial crisis, lending restrictions have been put in place (ranging from 75% to 80% LTV). As with international lending, the property purchased with the use of a loan is then done in exchange for a regular payment for the duration of the loan (known as the amortization schedule). The borrower also pledges that the property is set as collateral and if there is a default in the mortgage payments, then the lender closes the balance of the mortgage through a forced sale of the property. Any shortfall in the foreclosed amount and the outstanding loan amount can be pursued through relevant legal routes in the UAE. Key lending criteria include:

- First property – the borrower must put down a minimum deposit (this was 25%, but reduced in March 2020 to 20%).
- Subsequent properties – the borrower must put down a minimum deposit of 40%.
- Expat borrowers are only permitted to borrow a total amount (principal and interest) of ×7 their annual salary.
- In Dubai, the law actually only requires that your debt payments come to no more than 50% of your income, so this will further impact the level of the loan you can take on depending on personal circumstances (and other debt obligations).
- Overseas buyers typically represent more risk to the bank, therefore, they will be typically asked to put down a higher deposit.
- Banks had typically only offered a mortgage up to the age of 65 years old (Expat) and 70 years old (UAE national), though recently relaxed this age restriction with the introduction of UAE retirement visas perhaps driving the need for mortgages permitted beyond retirement age. Mortgages here are evaluated on a case-by-case basis.
- A tightening of regulation in 2008 has seen borrowers requiring "mortgage life insurance" which is approximately 0.05% of the loan amount per month (fixed at the start of the annual cycle).
- A mortgage borrower is also typically expected to take out some form of property insurance, either independently or via the lending bank. This monthly payment is added to the service mortgage payment as an additional cost.
- Banks also typically carry out a stress test on interest rate fluctuations. Currently, this might range from 3% to 8% and this assesses the affordability of the loan if interest rates were to rise.

Prepayment of mortgages in Dubai

It is common for mortgagors to be permitted to pay off their loan amount before maturity. An annual pre-payment amount is typically permitted without charges (normally 10%–20% per annum). Beyond that amount, a percentage of the loan amount paid is typically charged. The current UAE Central Bank legislation stipulates that this is 1% or AED10,000 whichever is lower. Mortgage buyouts from another bank can be charged at higher rates as much as 3%. The maximum term of a mortgage loan in Dubai is 25 years (age-permitting).

Calculating the full cost of borrowing in Dubai

Let us assume an expat takes out a mortgage in Dubai for the purchase of an AED3 million villa. His associated costs to this purchase would be:

- A minimum 20% down payment, AED600,000, leaving AED2,400,000 as the loan amount.
- The mortgage processing fee of 0.5%–1% of the loan amount (although can be capped by some banks), AED12,000–AED24,000 (let's assume it is capped at AED15,000, with 5% VAT making it AED15,750).
- Valuation fee (+5% VAT): AED3,150 (instructed by the bank).

In total, this would mean they would need to pay AED618,900 to secure the mortgage finance, represented as a 20% deposit plus additional 0.63% in bank fees. In addition, the purchaser would also need to pay the agent fee (2% + 5% VAT) and DLD transfer fee (4%) for the purchase in the secondary market. This a further AED183,000 to his property purchase (6.1% of the purchase price). Alongside these more well-known costs, there are some additional transfer costs at the Lands Department that include:

- DLD transfer admin fee of AED580
- Trustee registration fee of AED4,200
- Mortgage registration fee at the Lands Department equal to 0.25% of the loan amount plus AED280 admin fee, AED6,280

In this example, the expat's upfront cost to purchase an AED3 million villa in Dubai would be AED812,960. The transaction fees (AED212,960) represents a total purchaser cost of 7.1%. In comparison to other global markets, this figure appears reasonable against UK (10%); US (9%); Australia (12.5%), as suitable comparables.

Cash (unleveraged) vs mortgage (leveraged) purchase analysis

The earlier example I gave in Chapter 2 showed the impact of leverage when evaluating a property purchase. The leveraged case gave the investor a much higher IRR (15% vs 8%). However, there are other considerations to make when taking on a mortgage in Dubai. And of course, leverage provides stronger returns in a rising market, however, in falling markets it exposes the investor to some inherent risks. Banks in Dubai generally require mortgagees to repay interest + principle with each loan repayment. This would be recognized as a "conventional mortgage". This would mean that each monthly payment is equivalent to the amortization of the loan for the specific term agreed at inception of the loan. Typically, mortgages are taken for 25-year terms in Dubai, but as with other markets, they can be shorter depending on the personal circumstances of the borrower. Interest-only mortgages do not currently exist in the market. Whilst the term can be for as long as 25 years, the mortgage is typically only fixed for 1–5 years and beyond this, it would switch to a variable rate (typically based on a fixed margin + 3M EIBOR). Web-based mortgage calculators are typically available to check payment schedules on a mortgage. However, one can also apply the annuity formula for a more specific analysis of their mortgage schedule. This formula is based off a present value annuity calculation:

$$= [1-(1+i)^{-n}]/i$$

where:
i = interest rate (%)
n = loan period

From this number, one can take its reciprocal to get the mortgage payment multiplier. This number can then be multiplied against the loan amount to determine the annual mortgage payment. This can be subsequently used for monthly payments. While this payment amount stays constant (if interest rates stay constant) for the loan duration, as the debt is repaid, more of the payment amount goes to pay off the loan principal. In a longer-term mortgage, say 20 years, a larger portion of the borrower's mortgage payment will be used to pay interest rather than the principal in the earlier years (up to Year 7). If the loan term was 10 years rather than 20, the borrower's mortgage payments will pay off a larger proportion of the principal versus interest. Of course, on a shorter-loan term, the monthly payments are significantly higher and run the risk of a default if the borrower was to lose their job or rental income. Therefore, it might be pragmatic to keep the monthly payment to an amount that is at least equivalent to the rent that would be received on the unit.

Let's take an example, a one-bedroom apartment in Dubai Marina was recently purchased for AED1,000,000 using a 20% deposit (AED200,000) and an 80% loan (AED800,000) on a 3% interest rate. The equivalent unit currently rents for AED80,000 per annum.

If the purchaser, was to take on a 25-year term, the annual mortgage payments would be AED45,942 per annum:

Interest	3%
Loan term	25
Annuity (PV multiplier)	17.41315
Mortgage payment multiplier	0.057428
Loan amount	800,000
Annual mortgage payment	45942.30

If we apply the advice of keeping the mortgage payments to an equivalent market rental payment (inclusive of annual service fees), we can use the goal seek function in Excel to find the shortest loan term this purchaser would be sensible to take (to maintain rental coverage). The outputs from this analysis below show that the purchaser would be sensible to take a loan no shorter than 14 years (see below).

Apartment size (sq.ft)	715
Service fee per sq ft (in AED)	15
Annual service fee (in AED)	10,725
Gross rent (in AED)	80,000
Net rent (in AED)	69,275
Goal seek with annual mortgage @3% per annum to be AED69,275 (fixed) to find loan term (years)	14.39 years

Interest	*0.25%*
Loan term	300
Annuity (PV multiplier)	210.8765
Mortgage payment multiplier	0.004742
Loan amount	800,000
Monthly mortgage payment	3,793.69

To apply this same technique to match many of the online mortgage calculators whose output is typically shown as a monthly payment, then the same present value annuity formula above can be used, however, the interest rate would need to be expressed as a monthly rate, in this example, that would be 0.25% (3/12) and the loan term would become 300 months (12 × 25). This gives a monthly payment of AED3,793.69. The table above shows the outputs from applying this revised monthly information to the same loan terms.

Conclusions

This chapter has provided an overview of Dubai's residential property cycle and evaluated the key drivers over the last 20 years, since foreign ownership was permitted in Dubai. When studying cycles, we are trying to understand the causal effect linking these to a current pricing cycle. The danger with cycles is the suggestion that they are replicated and can be predicted. Often this is not the case. A more suitable approach to studying cycles is to evaluate lessons or outcomes to help improve market and future investor confidence. The case study "a tale of two expats" brought to the front of this evaluation the impact of too much leverage. Highly geared markets tend to promote speculation and from the study of the Dubai residential market, accounting to some extent for the high level of volatility in the early cycles. Government policy and stricter lending criteria put in place since 2014 have seemingly tamed the market deviations. Yet, market maturity (a longer-time series to base decisions) as well as better data (see Chapter 8) has also attributed to less volatile price movements. Long-term fundamentals and pricing should be more consistent in the future, especially with the imminent intro-duction of UAE Treasury Bonds (see Chapter 5). As risk is managed and more foreseeable in the market, the uptake of mortgages is likely to increase, and this will also create greater confidence in both bank lending (at lower rates) as well as buyer (particularly end-user) demand.

The next chapter moves these discussions into understanding how Dubai residential can add diversification benefits to a global property portfolio. The content of this chapter has served as a suitable starting point for this analysis. Chapter 4 takes more of an appraisal outlook, particularly useful for those looking to invest into Dubai residential property.

References

Baum, A. (2022) *Real Estate Investment: A Strategic Approach* (4th ed.), London: Routledge.

Colliers (2009) *Colliers International House Price Index: Dubai Q1 2009* (printed), Dubai: Colliers International.

Data Finder (2021) https://www.datafinder.ae/ (Accessed March 2021).

Deloitte (2021) Middle East Real Estate Predictions: Dubai.

Dubai Statistics Centre (DSC) (2021) https://www.dsc.gov.ae/en-us/Pages/default.aspx.

DXBInteract (2022) https://dxbinteract.com/.

Knight Frank (2022) https://www.knightfrank.ae/research/dubais-covid-comeback-2022-8855.aspx.

Mortgage Finder (2019) 'Islamic and conventional mortgages explained' https://www.mortgagefinder.ae/blog/islamic-and-conventional-mortgages-explained/.

Property Monitor (2022) https://www.propertymonitor.ae/.

Waters, M. (2019) A critical examination of property valuation variance in Dubai, PhD thesis.

4 Building a global residential portfolio with Dubai real estate

The main reasons for investing in real estate outside a home market would be for diversification and higher returns (Baum and Hartzell, 2012). However, these apparent benefits do not come without risk and uncertainty. Studies have typically found the security of property rights; expected returns; liquidity and market size as key determinants impacting the choice of real estate investments. Availability of market information and performance benchmarking are also key criteria. As with the commercial markets and fund managers, individual investors now see it valid to spread their residential property investments across different geographies. With the current period (2010–2022) of low interest rates and financial liquidity seeing a convergence of property yields in core markets, investors are naturally seeking better returns in new markets like Dubai, albeit still a relatively small proportion of global investment. The two core reasons for international investment: diversification and higher returns, in the context of Dubai real estate, should be our starting point.

The main objective of portfolio-based investment is to maximize returns and minimize risk, via diversification; be it investing in different asset classes; different geographies and as an expat in particular, in different currencies. The main premise to this is to invest in assets that are not perfectly, positively correlated. In real estate, this is best achieved via geographical diversification. So how effective is investing in Dubai residential property as a diversifier on an individual's global wealth portfolio? The following sections examine why Dubai and what things to consider when making a residential purchase. Below is a range of drivers and barriers to investing in Dubai.

Key drivers of property investment in Dubai

- Dubai provides property investors with a long-term residence "golden visa" (for ten years renewable for property investments above AED2 million).
- Investment returns are tax-free (both rent and capital gains) providing international investors the opportunity for higher real returns

DOI: 10.1201/9781003186908-6

Box 4.1 The why, how, what and when of diversification with Dubai residential real estate

The main reason for diversification is to ensure a portfolio is not exposed to a high number of similar assets (by class; location; currency) that would ultimately move synchronous with one another during economic and business cycles. Real estate has a negative or low correlation with stocks and bonds, proving its value in a mixed portfolio, by providing a more "efficient frontier". The main premise is that the inclusion of real estate should either offer an investor the same return at a lower volatility or a higher return for the same amount of risk. In terms of Dubai residential I advocate its position as offering investors higher returns. Recent global studies suggest at least 20% of our portfolio should be allocated to real estate.

Gross yields in Dubai are 2-3% higher than many other global core markets on a traditional annual lease basis. Residential investors also can leverage the booming tourist and short-stay business income stream, which could boost returns further, typically by as much as 40%, moving the yield into a double-digit return (Arabian Business, 2021). The flip side of the higher return is higher risk. However, there is no valid reason to expect Dubai's market to become more risky as time progresses. We are already seeing the market becoming more transparent and more liquid.

More recent studies have suggested diversification should be more focused towards residential property than commercial property, as residential property is less linked to the overall performance of the economy (Qayyum and Khan, 2021). Direct real estate suffers from being a large capital investment and so retail investors in particular are limited in scope to diversify through direct purchase of residential units in the absence of large capital sums. However, Dubai is now offering investors fractional ownership either via DLD title registration, permitting up to four owners per property, and crowdfunding via third-party investment platforms (see Chapter 2).

The final part, when? In fact it does not matter – the purpose of portfolio diversification is to invest in the merits of the asset in the long-term. The good news now is that we have much better data in Dubai to inform more rational decision making and reduce the volatility historically seen in the market (alongside leverage controls referred in the previous chapter). As time moves on, a much greater number of studies will start to look at the historical risk and return profile of Dubai as it has done in many other global markets.

- Dubai has a US-pegged currency (AED), therefore offers international investors a long-term security of investing in a leading global currency (USD).
- Annual rate of returns is higher than many other key global markets: Investors have been attracted to Dubai as it still offers attractive yields on the investment. An average residential yield in Dubai is c. 6%–7%. When compared to other global cities this is highly attractive. The average prime residential yield across the 30 cities stands at 4.5% in December 2021 (Savills, 2022), whereas for Dubai prime residential it sits at 4.75%.
- Annual rates of return are typically higher in Dubai as rent is collected in advance (annually, bi-annually or quarterly) boosting the real return for investors (when compared to monthly rent collection).

Key barriers to property investment in Dubai

- Home bias: A range of academic studies have highlighted the fact that in countries where foreign workers make up large proportions of the workforce, investment decisions are often tied to their home country. Studies found that in locations where expat workers felt "temporary" or an "uncertain" period led to an increased propensity to making financial commitments/investments in their home country. Fisher and Jaffe (2002) also found that an affordability criterion (arbitrage) exists between their host and home country. The study also found that home ownership in the host country was based upon a relativity of renting vs owning and other credit constraints.
- Affordability: Expats often refer to the relatively high LTV deposit required when purchasing a property with a mortgage in Dubai. Currently, the requirement is for a buyer to commit at least 20% of the purchase as a deposit. This may appear high when compared to the 5%–10% required in their home market. In fact, on other metrics such as % of monthly income on rent or equivalent mortgage, Dubai appears "affordable" (see below).

Home bias in investing is somewhat counter-intuitive to the rationale for diversification. The high LTV, whilst a barrier to entry, has proved instrumental in curtailing price speculation and volatility (discussed in Chapter 3). Further still, affordability in investment is always one of relativity. Whilst UBS (2021) called out Dubai as "undervalued", many residents still perceive it as expensive and unaffordable. Table 4.1 highlights the relative change in affordability in Dubai's residential sector over the last ten years, with the equity contribution (−23.8% less) and monthly repayments AED1,000 per month higher in 2021 than they were in 2011 despite the noticeable 33% increase in property prices. More importantly, based on a single (or dual income of AED25,000 per month) and a single (or dual income of AED50,000

Table 4.1 Housing affordability in Dubai (average price, 2,000 sq ft)

Purchase date	Income	Deposit	Purchase price	Deposit	Loan amount	Interest rate (%)	Mortgage repayment per month[a]	Mortgage as a % of monthly income
May 2011	25,000	35	1,500,000	525,000	975,000	6.5	6,583	26.3
May 2021	25,000	20	2,000,000	400,000	1,600,000	3	7,587	30.3
	50,000[b]	20	2,000,000	400,000	1,600,000	3	7,587	15.2

Source: Author's own.
a Assumes a 25-year conventional mortgage.
b Assumes a dual income per person per month.

Table 4.2 Saving for the mortgage deposit

Saving for the deposit				No interest	2% interest	5% interest	8% interest
25,000	20%	5,000	400,000	80	75	69	64
25,000	25%	6,250	400,000	64	61	57	53
25,000	35%	8,750	400,000	46	44	42	40
50,000	20%	10,000	400,000	40	39	37	35
50,000	25%	12,500	400,000	32	31	30	29
50,000	35%	17,500	400,000	23	22.5	22	21.5

Source: Author's own, based on assumed income; saving propensity and number of months.

per month), the mortgage repayments to purchase an average 2,000 sq ft property today sit comfortably within the income affordability of 30%–50% that is stated globally (a measure of the proportion of income spent on housing costs).

The counter to mortgage affordability is the 20% lump sum deposit required to purchase. Table 4.2 presents a case on how an expat can save towards their deposit, where they should keep the funds and how long it should take to save for a 20% down payment, based on an assumption that they will look to purchase the same AED2 million property.

Table 4.2 shows an analysis of the same income assumptions versus a % propensity to save for the deposit and an analysis of how many months would it take for that individual or household to save for the deposit. The analysis shows that it is the % amount saved per month that impacts the time it would take to save for the deposit. Marginal differences are made when investing the deposit savings in different financial products, assuming 2% (low risk); 5% (moderate risk) and 8% (high risk) per annum returns. The information shows that AED25,000 per month household income can save a deposit for an average residential property within 3–5 years. An AED50,000 per month household could reduce this to less than two years if regularly saving 35% of the monthly income. This information provides a useful plan for those looking to save for a deposit and then invest in a property in Dubai. Of course, if the property requirements were less or more then there would be some variance to these figures. For example, an investment property of AED1,000,000 (c. 900 sq ft) would take Household A (AED25,000 pm), between 20 and 24 months to adequately save for a deposit.

The next section moves on to why real estate should be part of your investment portfolio.

How does investing in real estate compare to other asset classes?

Global studies have shown how attractive real estate investing is to a personal investment portfolio. For Dubai, the time series for a thorough analysis is too short, so let us first look at what the data are saying for a more mature real estate market. I take the example of my home country, the UK. Over the last 50 years, empirical research (Baum, 2022) has found that UK real estate offers investors a higher return per unit of risk (0.95 versus 0.45 for equities) Similar results were noted in the US (1.23 versus 0.87). The analysis shows real estate performs relatively well against stock and bond markets on a risk-weighted return basis. If long-term returns are similar, then this implies real estate markets are less volatile. Key reasons for this are the quasi-investment characteristics of real estate between bonds and equities; the financial lever-age; faster pace of transactional activity; and more speculative behaviours that runs through an equity market versus the real estate markets. The significant cost to purchase and trade real estate typically prohibits quick and reckless decision-making.

Global studies have shown that investing in direct real estate provides the most optimum return per unit of risk over a long-term investment horizon, marginally lower annualized returns versus the stock market, yet significantly lower levels of volatility as a proxy for risk. Of course, over other time periods or longer time series, direct property is unable to consistently out-perform the stock market, but there is clear support that real estate can enhance returns within a stock and bond portfolio. Does Dubai show similar patterns, despite its relatively short time frame for analysis? The next section examines why Dubai might be a good diversifier for an expat or foreign investor's investment portfolio.

Why Dubai might help diversify your property portfolio?

Candelon et al. (2021) found that international diversification in a real estate portfolio was advantageous as the assets have a lower correlation than in a single geography (economic, political, tourism) therefore providing a lower risk to the investor's return based on a single negative event in Country A will not be as impactful if assets are also held in Countries B, C and D. In this context, Dubai has benefitted from global investment, notably more so at the private investor/HNWI level (and largely in residential assets). Foreign investors from 2002 onwards have certainly had ample opportunities to make financial gain in Dubai's residential property market (see Chapter 3). Table 4.3 shows Dubai house prices performed against a basket of foreign currencies. In terms of AED invested, Dubai residential gave investors a 2.1% annualized return (over 10 years) and 3.7% pa (over 20 years). If we add this to a historic rental of c. 7%–10% pa, Dubai residential has offered very strong yields for those seeking long-term growth. Over the long term, international property investors have benefited from not only capital price growth in the Dubai residential market (AED return = %), the continued strength of the US-pegged currency versus a basket of major global currencies further exacerbated returns for an investor when repatriating those funds back to their home country. For example, a British expat investing in

Table 4.3 10-year/20-year returns based on a basket of major global currencies

Return	AED (%)	Euro (%)	Australian Dollar (%)	Chinese Yuan (%)	Swiss Franc (%)	Indian Rupee (%)	Pakistani Rupee (%)	Russian Ruble (%)	British Pound (%)
10-year	23	50	75	23	35	104	104	222	43
20-year	105	81	54	60	37	217	217	369	139

Source: Supplied by Taimur Khan.

Dubai residential would have made a 43% return (10-year) or 139% return (20-year) period based on the favourable currency movements during those periods (equivalent to 3.6% pa and 4.5% pa, respectively). Russian investors would have made remarkable returns of 222% over a 10-year period (8.3% pa) and 369% over a 20-year period (6.75% pa) based on the currency depreciation seen in the Ruble. However, trends like this are periodic and currency movements as stated previously do impact the potential for real returns to be either improved or eroded, depending upon the situation of the investor's home currency over a specific period. What does stand out though is the ability of Dubai residential to offer investors significant annual total returns.

In the first wave of property market activity and significant price growth (2002–2007), returns were double-digit and lending on off-plan purchases was largely unregulated and unsecured. Personal loans and even credit cards were used to raise the 10% deposits on multiple off-plan units, to then sell on at a high profit, was a common occurrence. In 2007/2008, the Economist observed in Dubai the presence of negative real estate rates (inflation greater than cost of borrowing) driving up real estate prices, making a strong investment case for borrowing from banks and investing it into real estate. We may see similar macroeconomic conditions in 2022 and beyond.

Global studies by MSCI have shown that there is a lack of correlation between real estate versus stocks and bonds (either low positive or negative correlations). Studies in the UK and US both show that property has a low correlation to bonds (0.03) and the stock market (0.27). This means that it is likely that the real estate market and equities are driven by a different range of factors, making real estate good for diversification in a mixed asset portfolio. So, what do we see in the Dubai markets?

According to Roche (2019), it might be expected that real estate and equities are evidently much more closely correlated with local stock markets than seen globally. The paper argued that real estate companies in Dubai make up a significant part of the DFM (Dubai-bourse) market capitalization, therefore movements in equities and movements in real estate will be more synchronous.. For Dubai, the overall correlation between property and equities, since January 2016, is +0.68, proving that investing in both assets will do little to diversify your investment portfolio. As with the Western studies, Dubai equities run at higher levels of volatility, with Roche (2019) citing this volatility to greater speculation in the equities markets in Dubai. As the DFM introduces more sectors and IPOs in more varied economic sectors then one could assume this high positive correlation between equities and real estate will reduce overtime. This analysis would show that since 2016–2019 (a short time period), investors have not been able to achieve diversification between real estate and equities in Dubai. So if that is the case, which asset class should investors put their money? The next section examines this question further.

How do investing in equities and real estate in Dubai compare?

Despite Dubai being open to foreign investors since 2002, the time series to make a similar analysis is somewhat limited. Many residential property indices only provide verifiable data from 2013 onwards and the new commercial property index starts from 2016. Nonetheless, some suitable commentary does exist. A previous analysis done by Colliers in 2014 showed that during a seven-year hold period (2007–2014), an investment in Dubai residential property would have outperformed that of a Dubai equities portfolio. The analysis showed a 65% gain in property versus a 10% gain in equities during the study period. As with other global comparisons, the stock market also displayed much higher levels of volatility than that compared to residential property.

For expatriates in Dubai, a common question that is asked is: "Where should I invest?"; "Am I better off investing in a global stock/bond portfolio?"; "Investing in Dubai is too risky?". Extending the analysis above to fit that expatriate investment dilemma, over a ten-year hold period (backtesting 2012–2022), investing in Dubai residential real estate would have most often performed better than an investment into global stocks and bonds. Table 4.4 takes average sale price information across six popular residential freehold communities and shows that the majority of these outperform a 60/40 or 70/30 global stock:bond portfolio. I am not advocating a one or the other approach, yet the data provide strong evidence supporting the presence of Dubai residential real estate in their personal investing and furthermore often yields higher price gains. This analysis omits the returns from annual rent or dividends (for simplicity), but it would be expected that these are operating to a similar yield regardless of the asset. With rent returns included (at a long-term average of 8% per annum), Dubai residential would have offered an investor an annualized total return of up to 12.5% between 2012 and 2022. This should appear very attractive when compared to other global examples.

From these analyses, investing in real estate in Dubai looks like a credible option. But what should investors look out for when investing in a foreign market like Dubai? One key area in international investing is the management of currency risk. With Dubai being pegged against a global safe-haven currency like the US Dollar, Dubai appears an attractive proposition for investors. The next section provides some discussion about managing currency risk and concludes by examining the performance of Dubai residential versus a basket of major global currencies.

Managing currency risk when property investing in Dubai

The impact of currency movement has a large impact on the process of property investment pricing and portfolio construction (Baum and Hartzell, 2012). The academic community have had a mixed debate on the suitability of currency hedging with real estate investing. With fluctuating rates, the value

Table 4.4 Dubai (Residential) vs Global Stock/Bond Portfolio (ten-year return).

	Dubai – Residential (Ready) Price Per Sq Ft, Sample Freehold Communities						Global Stock/Bond Portfolio		
	Business Bay	JLT	Jumeirah Village Circle	Downtown Dubai	Dubai Marina	Palm Jumeirah	60/40	70/30	100
10-year return	47.15%	24.28%	9.95%	38.02%	55.48%	50.56%	22.35%	36.20%	182.26%
Annualised return	3.94%	2.20%	0.95%	3.27%	4.51%	4.18%	2.04%	3.14%	10.93%

Source: Data sourced from DXBInteract.com and Portfolio Visualizer. The data above is capital gains only and does not evaluate the annual rental income or reinvested dividends.

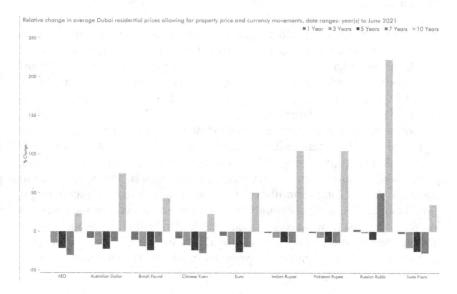

Figure 4.1 Average residential prices vs major global currencies.
Source: Supplied by Taimur Khan.

of a successful investment property could be devastated when converted to the investor's currency (Worzala, 1994). This risk should be recognized and incorporated into the international investment decision; but, as found by the research of Worzala (1994), the end-goal is to examine real estate returns in home market currency rather than base a decision on the moving exchange rate. By investing in a range of global currencies, the investor is somewhat protected from any wild swings.The author here and similarly others have stated the inherent mismatch between the holding periods of real estate versus currency hedging tools, does make managing currency risk challenging. One might then take a preferred view to invest in international real estate backed by more stable global currencies. Moreover, the Dubai real estate market is effectively pegged to the US Dollar, and this will no doubt be the long-term currency of an expat worker in Dubai investing in real estate. To highlight the impact of currency movements, Figure 4.1 illustrates the price changes in Dubai versus some key currency baskets, demonstrating how currency movements affect investment performance. From these data, we can see that over a ten-year hold period most major currencies have performed positively when invested in Dubai's residential market.

Conclusions

This chapter has sought to bring a series of practical considerations together for those evaluating residential property investments in Dubai. Real estate as an asset class does have some tangible benefits that make it a suitably attractive investment choice. The information provided data-driven commentary on the relative performance of real estate against other major asset classes, both globally as well as in Dubai. The long-term investment return profile of Dubai has offered investors both high yields and periodic opportunity for high capital gains, yet the inherent market uncertainty was clearly priced in during its early years (2002–2011). This was reflected through higher discount rates/lower cap rates (multipliers). These yields have compressed over time and uncertainty dissipated as the market became more regulated and internationalized. International investors face core challenges relating to managing risk and uncertainty throughout the holding period, and the most apparent is currency risk (which is notoriously hard to manage). Yet, the long-term 10- to 20-year returns, against a basket of major global currencies, have seemingly boosted the actual returns for investors in those markets, proving to an extent that Dubai residential property does have a greater role to play in an individual's portfolio. The next chapter turns to evaluate the commercial property market in Dubai.

References

Arabian Business (2021) 'Dubai Real Estate Short Term Lets, Long-term rewards'. https://www.arabianbusiness.com/industries/real-estate/465822-dubai-real-estate-short-term-lets-long-term-rewards. Accessed December 2021.

Baum, A., and Hartzell, D. (2012) *Global Property Investment: Strategies, Structures, Decisions*, Chichester: Wiley-Blackwell.

Baum, A. (2022) Real Estate Investment: A Strategic Approach (4th edition), London: Routledge.

Candelon, B., Fuerst, F., and Hasse, J. (2021) 'Diversification potential in real estate portfolios', *International Economics*, 166, pp. 126–139.

Colliers (2014) DFM/HPI comparison investment performance:, Dubai House Price Index, Q1 2014 (printed), Dubai: Colliers International.

Fisher, L. & Jaffe, A. (2003). Determinants of international home ownership rates, Housing Finance International. 18, pp.34-42.

Portfolio Visualizer. https://www.portfoliovisualizer.com/.

Qayyum, A and Khan, W.A (2021) Impact of global residential real estate on portfolio diversification, Journal of Real Estate Portfolio Management, 27 (2), pp. 149-165.

Roche, J. (2019) Real Estate Markets and Stock Markets: Two Stories, or Just One?, Cavendish Maxwell. https://cavendishmaxwell.com/insights/opinion/real-estate-markets-and-stock-markets-two-stories-or-just-one.

Savills (2022) 'Savills Prime Residential Index: World Cities', World Research – February 2022
https://pdf.euro.savills.co.uk/global-research/savills-prime-residential-index---world-cities-february-2022.pdf

UBS (2021) UBS Global Real Estate Bubble Index 2021. https://www.ubs.com/global/en/wealth-management/insights/2021/global-real-estate-bubble-index.html

Worzala (1994) 'Overseas property investments: How are they perceived by institutional investor', *Journal of Property Investment*, 12, pp. 31–47.

5   Commercial

In Chapter 2, an overview of commercial property market activity was discussed. This chapter will expand on commercial market practices in Dubai. The focus of this chapter will be:

- Overview of commercial property as an investment
- Main participants in the investment market
- Routes to investing in Dubai's commercial property market
- Drivers of property returns
- Future outlook for Dubai's institutional market

Commercial property as an investment

The income and risk characteristics of direct property vary and these will depend on lease conditions – the shorter the length of the lease or the period between rent reviews then the more quickly rents will adjust to the real economy and to inflation. In Dubai, the average length of a lease is now around 3–4 years (but can be as short as 1 year), depending on the sector. In other markets, we see much longer lease lengths, and long leases in property mean its income characteristics behave more like bonds. In Dubai, with shorter leases, income from real estate exhibits more volatility and these equity-like characteristics are greater during periods of strong economic growth and rising rents. This does not mean Dubai real estate is correlated with the stock market (though this was noted in Chapter 4), it just means it exhibits similar characteristics to a stock market and is more prone to shocks derived from shorter reversions to market conditions (apparent from the market practice of shorter leases). In fact, global studies show property is counter-cyclical to bonds and equities and this supports the presence of real estate in a well-diversified investment portfolio. Yet institutions (pension funds) have been historically hesitant to invest in Dubai's commercial property market. Box 5.1, later in this discussion, explores why commercial property has yet to attract the attention of global property funds.

Table 5.1 provides an overview of the investment qualities behind the main asset classes available for investors in Dubai. One key difference

DOI: 10.1201/9781003186908-7

Table 5.1 Investment qualities of commercial property in Dubai versus equities
and bonds

	Equities (DFM)	*Bonds*	*Commercial property (Dubai)*
Income return	Variable, dividends are not contracted	Fixed	Fixed, short leases, though contracted agreements[a]
Capital gain	Variable, volatile	Fixed	Variable, lower volatility than DFM
Liquidity	Liquid	Liquid	Illiquid
Links to economy	Linked to economy/ inflation. DFM had been weighted heavily to financial and real estate sector. More IPOs being issued in other parts of the economy	Depends on inflation	Linked to economy. Inflation depending on rent reviews. Rents are linked to a rental index (or fixed increases)

Source: Based on Hoesli and MacGregor (2000), commentary provided by author.
a It is also important to note that in some instances in the offshore free zones such as TECOM
   and DIFC, the "landlords" have signed master agreements with the regulated authorities
   and as such rents must be in line with their current levels and potentially not as flexible as
   could be achieved in the open market.

in Dubai versus other global locations is the relatively short lease durations that exist, historically a by-product of the short length of commercial trade licences. This prevents commercial property benefitting from a
secure long-term income, which we might see in mainstay global investment hubs. Instead of commercial leasing practices in Dubai offering investors a long-term, inflation-linked income return, the short-term nature
of leases means there would be a higher holding risk, as future rents will
be tied to the economy and wider market movements. Furthermore, rent
increases are based on the Dubai Land Department (DLD) rental index,
prohibiting landlords from raising rents beyond the index (though fixed increases do exist, see Chapter 6). Within the commercial property market,
different asset classes will perform differently based on their specific value
drivers. For example, office property is driven by the economy (occupier
demand); levels of new supply (competition) and obsolescence; retail value
is driven by catchment demographics (purchasing power); competition;
tenant quality/mix (attractiveness); and trends in e-commerce; and industrial property value is driven by distribution; accessibility to major
transport nodes; as well as trends in e-commerce. It is also important to
mention that in the newer commercial zones such as TECOM and Dubai
South where global occupiers may consider bespoke commercial facilities,
the underlying land ownership structures (being ground leases or development leases) add a further layer of perceived risk when being considered
by foreign capital.

Property transactions in Dubai's commercial market

Transaction data for Dubai's commercial property market are separated into three distinct sectors: office, retail and commercial development land. As with the residential sector (in Chapter 3), commercial property has been cyclical. Table 5.2 presents the price changes in each sector from 2008 to 2021, extracted from Data Finder's database. The information presented is the average AED/sq ft. The trend line from this dataset broadly follows the cyclical patterns of the residential market, with the notable exception that the office sector seemed to demonstrate some resilience to the impact of the Global Financial Crisis (GFC) with average prices increasing steadily from 2008 to 2014. Beyond 2014, all sectors tracked the general Dubai line of a six-year period of gradual price declines. Like any city, office prices and occupancies are heavily location-specific, again, the regulatory environment plays a significant role in this regard as, for instance, those regulated by Dubai Financial Services Authority/Dubai International Financial Centre (DIFC) must have their corporate offices within the DIFC where there has been a shortage of available offices which has impacted the underlying dynamics whereon those occupiers who are governed by Dubai Economic Department have a much wider supply at almost all conceivable price and quality points.

Figure 5.1 maps out these cyclical patterns as a total commercial price change, illustrating some of the market volatility observed in Dubai's commercial property market since 2008. During the midst of the GFC, the frenzied real estate development in Dubai from 2005 to mid-2008 resulted in a series of newly built towers standing empty (Mirza, 2009). Office occupancy stood strong at 98% in 2008, but drastically fell to around 80%, prompting a rethink about whether new office development would perform as strongly as it had done.

Figure 5.2 highlights the dominance of transactional volumes for smaller offices in 2008–2022 (500–1,500 sq ft), prompted by the strata ownership law that promoted investment in commercial office stock from local and Gulf Cooperation Council (GCC) retail investors. Since 2014, a higher number of transactions are taking place in small offices (<500 sq ft), which indicates the investment demand is now more evenly spread. The development pipeline and market performance are playing catch up to this shift in demand. A 2021 Gulf Business Report, highlighted the fact that commercial office space between 800 sq ft and 1,100 sq ft (8–10 staff) is oversupplied with much to choose from. If you have a company with 25–30 people requiring an office of between 2,500 sq ft and 3,500 sq ft, there is short supply and demand in attractive locations is high.

Key findings from the Q3 2021 Commercial Index release and current market performance for Dubai's commercial sector (DLD, 2021) include:

- Approximately 2,845 commercial property transactions were recorded in Q3 2021 (totalling AED33.9 billion)

Table 5.2 Dubai commercial price change (AED per square foot, by year): office; retail; land

Type	2008	2009	2010	2011	2012	2013	2014	2015	2016	2017	2018	2019	2020	2021
Office	621	725	780	904	908	913	1,114	1,102	1,000	1,000	849	698	650	670
Retail	793	700	658	538	498	691	1,000	1,133	857	997	804	900	773	1,100
Land		267			180	375	801	444	349	458	350	401	240	249

Source: Extracted from Data Finder, accessed in 2021.

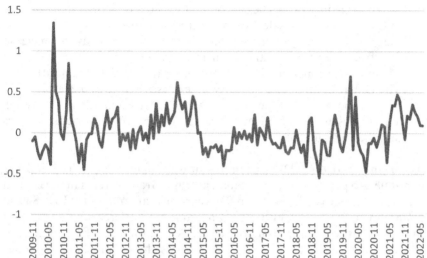

Figure 5.1 Dubai office property cycle (2009–2022).
Source: Data Finder (Accessed June 2022)

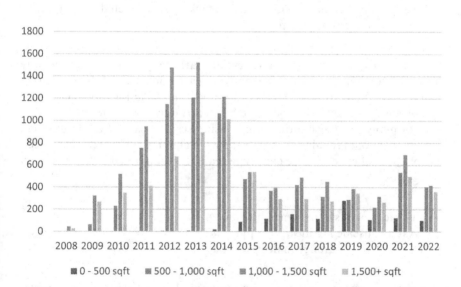

Figure 5.2 Commercial transaction volume, by property size (2008–2022).
Source: Data Finder (Accessed June 2022)

- Transaction volumes were 34% higher than in the previous 12 months. Ready (existing) properties accounted for 86% of transaction volumes in Q3 2021. Off-plan sales comprised the remaining 14%. This supports the viewpoint that buyers are looking to acquire a ready-income/cash flow from the premise rather than await a handover.
- There was a propensity for cash purchases (rather than financed).
- The Dubai Commercial Property Index rose sharply in the last quarter of 2021, both in the office sector and retail (1.4 and 1.5, respectively). This represented a significant price gain compared to previous quarters and the long-run ten-year index average

According to JLL (2022), office rents in well-managed quality buildings continued to perform well. In Dubai, average Grade A rents in the central business district (CBD) were up 9% year-on-year to about AED1,840/sq m/annum in 2022. However, the office market is tiered and segmented in performance, as referred to earlier. A snapshot of the commercial property market performance in 2022 across the major sectors includes (JLL, 2022):

- Office: 9.1 million sq m GLA (total stock); +1% expected new supply in 2022; 9% rental growth pa
- Retail: 4.6 million sq m GLA (total stock); +7% expected new supply in 2022; 5% rental decline pa

As noted in Chapter 2, the commercial market represents a significantly lower proportion of transactional activity when compared to the residential market. Yet there are some significant opportunities in the Dubai commercial property market and it ought to be an attractive proposition. The presence of many top-tier multinational businesses; the ease of doing business and international regulation supports such a claim. Yet the sector has faced historical challenges. These challenges can be summarized into three main categories:

- The legacy of strata ownership law
- Assets of scale are in short supply
- Lack of transparency in decision-making

Box 5.1 elaborates on these points further. Box 5.1 provides an industry insight from Faisal Durrani on the commercial property investment market in Dubai. It also draws on office transactions and explains why Dubai has yet to have seen a large investment uptake from institutional investors.

Box 5.1 Industry insights by Faisal Durrani, partner – Head of Middle East Research, Knight Frank

Dubai's commercial property investment market

Dubai's commercial office investment market has expanded rapidly over the course of the last 15 years (see Figure 5.3), with total transactional volumes crossing AED24 million during the first nine months of 2021, compared to just AED310 thousand in 2007. 2009, however, retains its position as the most active year for office deals, with over AED138 million worth of sales recorded during the height of Dubai's first property cycle and pre-GFC construction boom.

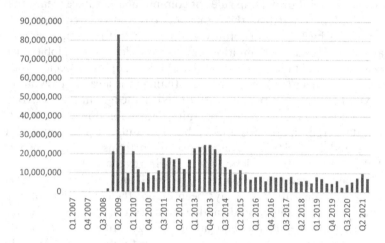

Figure 5.3 Value of commercial office transactions (in AED, freehold areas).
Source: Supplied by Faisal Durrani.

JLT and Business Bay dominate activity

Unsurprisingly, perhaps, the locations with the highest concentration of commercial office stock in the city – Business Bay (16 million sq ft) and Jumeirah Lake Towers (JLT) (11.8 million sq ft) – have and continue to dominate sales activity across the city's 110 million sq ft office market. During the heady days of Dubai's first market peak in 2008/2009, with prices in Business Bay crossing AED1,115 psf and JLT not far behind at AED740 psf, authorities opted to integrate elements of Australia's Strata Title system, to expand access to commercial office ownership. In essence, the law allowed for the subdivision of larger commercial floor plates, so that smaller investors were able to access the market at a time when rampant price growth threatened to stymie investment activity.

Key challenges in creating a truly institutional market:

Key issue 1 – The legacy of strata ownership laws

The resultant burst in transactional activity reflected the success of the introduction of strata-ownership in Dubai (dubbed the Jointly Owned Ownership Law); however, the legacy of that decision, since its inception in 2010, has had ramifications for city's office landscape that are still felt today. While smaller investors were able to access the office market, the joint ownership rules of communal areas have meant that many buildings, particularly in Business Bay and JLT remain mostly discounted by larger corporates, simply due to the complexities of having to negotiate with multiple landlords. Indeed, some global businesses, such as Standard Chartered and HSBC, opted to occupy their own headquarter (HQ) buildings less than a mile away in Downtown Dubai, driven in part by the strata-linked challenges in Business Bay. This however created an opportunity, albeit for domestic/regional capital wherein in both instances a build-to-suit was provided for the banks and the original development consortiums having exited in whole/part prior to completion (as below in key issue 2).

All have however not been lost. The small and medium enterprise business community has always been one of the core pillars of economic growth in Dubai and smaller office suites, such as those that have been spawned by the segregation of larger office plates into individual office units have and continue to cater to this cohort, albeit the supply-demand imbalance persists. Unsurprisingly, occupancy rates in submarkets like Business Bay remain stubbornly static at around 70%, against a wider city average of nearer 80%. However, rental growth too has lagged, with Business Bay and JLT experiencing rent increases of 12% and 10%, respectively, since 2012, against a city-wide average of 13%. Figure 5.4 highlights the changes in office rents across Dubai's sub-markets between 2016–2021.

Key issue 2 – Assets of scale in short supply

Another linked barrier to the emergence of a truly institutional investment market has been scale. The strata ownership rules in freehold investment areas like Business Bay and JLT mean that lot sizes are often well below the thresholds set by global institutions such as Blackstone, Blackrock or AXA.

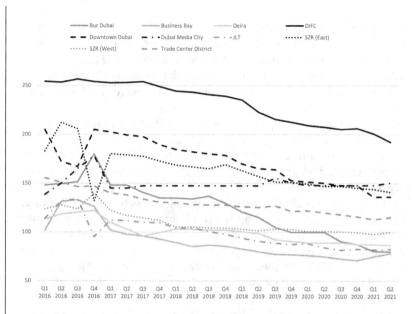

Figure 5.4 Performance of prime office lease rates – select Dubai sub-markets (AED/sq ft).
Source: Supplied by Faisal Durrani.

The lack of larger assets has been a difficult hurdle for the city to overcome, with only a handful of well-known commercial office buildings trading in the last 15 years, namely: Standard Chartered's Downtown Dubai HQ building, which includes tenants such as BP and DLA Piper, which was bought by the Kuwait Investment Authority (KIA) for AED 650 million (NIY 6.5%) in 2016 on a sale-and-leaseback arrangement and the 57-storey U-Bora Tower 1 in Business Bay, which was bought by Seynar Real Estate in 2018. Within the main Gate area of DIFC, arguably the most prime part of the development and financial CBD all the assets with the exception of two are owned by the DIFC themselves, the two assets both office properties are owned by Standard Chartered Bank (SCB, Asia) as their offshore HQ and another by Legatum the international boutique investment company – both assets having been acquired more than seven to ten years ago.

Key issue 3 – Transparency and lending lagging global counterparts

One of the other major barriers for institutional investors remains the lack of adequate transparency. Dubai leads the region in terms of market data availability, but it still falls short when compared to other global gateway cities. There is no central deals repository that is openly available and deal details are often kept under lock and key. This is in part due to cultural sensitivities of disclosing the full financials of major purchases. Primary research can always be relied upon to surface most details of such closed-door transactions, but it takes time.

Local and regional investors who are familiar with the city, its customs and regulations and the asset itself are far quicker to move on investment opportunities, often far quicker than global investors. This has perpetuated an apparent veil of market opaqueness for the global investment community.

One of the other challenges is the lack of access to local or regional debt. The significant growth in the residential markets has seen a majority of the bank lending being in this sector. The central bank has regimented caps on the lending that banks can avail to the real estate markets and this is not always segmented across sectors. The shortage of local conventional debt or at least debt on commercially attractive terms has also seen a rise in Islamic finance instruments. These are well understood by local and regional investors but often provide a further hurdle to international capital, especially with some of the potential restrictions that can accompany around potential tenants (i.e., would be a challenge to generate an income from a non-Shari'a tenant and this includes in many instances banks as they generate revenue from the charging of interest).

A "green" route to attracting institutions?

While these are some of the challenges that have held back the Middle East's most dynamic property market from going truly global, there is one rapidly emerging reason for optimism in the counterintuitive form of green-rated buildings.

With environmental, social and governance (ESG) considerations riding high in the minds of global institutions, the planet-wide hunt is on for green-rated assets as investors look to rebalance portfolios and off-load "brown" buildings.

The UAE is currently home to 869 green-rated buildings (550 of which are in Dubai), the 14th highest concentration globally and the only nation in the Middle East in the world's top 30. The US leads the league table with almost 81,000 green buildings and at the city level, London ranks first with 3,000 environmentally accredited buildings.

The ESG agenda has become mainstream, and we are already seeing evidence of its impact on rents and occupier behaviour in Dubai. Take the newly completed Supreme Court Chambers on Dubai Creek, for instance. The automated car parking system results in less vehicle emissions and makes more efficient use of space; a clear indication of how sustainability sits at the heart of the user experience and has been key to its success.

As Dubai's prime office stock in core location ages, the calls to refurbish and greenify will grow louder, or such assets could potentially face rising voids and potential obsolescence. This presents another tantalizing opportunity to draw in global investors as the carbon footprint of a refurbished building is lower than that of a newly completed one.

While smaller businesses may not yet be demanding green-rated buildings, institutions are, and therein lies Dubai's opportunity to capitalize on its already high number of green-rated assets. Developers and landlords in the Middle East have made great strides in delivering world-leading green buildings, incorporating stunning technologies such as the condensation harvesting system on the Burj Khalifa, but such technologies need to be adopted more widely if landlords are to compete on a level playing field for increasingly green-conscious businesses and investors.

Main participants in Dubai's commercial property market

The participants in Dubai's commercial property investment market have traditionally been locals and GCC investors, high net-worth individuals (HNWIs) and the government. Institutional investors, sovereign wealth funds and other international investors have typically opted to invest in commercial property assets in other global markets, perhaps for the benefit of geographic diversification, but also the investment characteristics in core global markets better match their investment goals or fund liabilities (secure and stable). Overseas investors do invest in Dubai's commercial sector, however, often favouring a commercial development fund, for example, where there is a specific target return requirement matched by the fund over shorter time periods (for instance, 30 months being the average time to design and build a tower block in Dubai (Mirza, 2009)). Such funds carry different risks but are on a 2–3 year time horizon with a clear exit strategy, rather than commercial property which carries a longer-term risk over 7–10 years. The majority of international capital has come into the market indirectly through investment in the underlying holding company or developer rather than at the asset level and usually against a wider collateral base, such as a share pledge.

Drivers of commercial property investment returns in Dubai

Commercial yields in Dubai

Commercial property yields have remained somewhat unchanged with only a slight compression in yields in the ten years between 2008 and 2018, with yields moving on average from 8.75% to 8%. That said, a few of the institutional-grade properties are trading at significantly lower yields. Table 5.3 illustrates the levels of yield compression over time (2010–2019), referencing a range of city-wide office transactions in Dubai. The cyclical pattern of commercial property values seems to have been moving in line with the yield movements. The lack of institutional-grade assets also means it is hard to extrapolate too much from the evidence below. What we have seen however is a willingness for funds to pay reasonably low yields for commercial assets in Dubai. Recently, in 2022, the market has observed sub-6% transactions (NIY, 5.80%) for a prime office building located in a core commercial free zone, indicating the appetite for top-quality Grade A assets. In the next 5–10 years, it would be expected to see strong quality assets trading well. For example, top Grade A institutional assets, like ICD Brookfield Place in DIFC. This is based on its high-quality institutional position (blue-chip multinational tenants, Grade A specifications, long leases 5–9 years). Box 5.2, later in this chapter, elaborates more on ICD Brookfield Place as an institutional case study. It provides a good example of the new types of commercial property development taking place in Dubai.

Table 5.4 provides yield evidence across a range of commercial assets in Dubai in 2018. Long-lease single-owned office buildings in Dubai with a strong tenant appeared to be in the range of 6.75%–7.25%, yet further yield compression has occurred since then as real estate serves to be an attractive long-term investment, particularly for local and international funds. As we move through the yield spectrum, we have seen more alternative investments like schools and healthcare facilities at higher yields of 8%. However, there has been notable interest in healthcare and education assets from Islamic funds as they are Shari'a compliant. The main issue is that the underlying land for these assets is often owned by the Government (and KHDA for most schools) and such leases are not always transferable. The spread on retail

Table 5.3 Prime yields based on specific institutional-grade transactions (offices)

Year	Property	Yield (NIY) (%)	Sales evidence
2010	Emaar Square, Downtown Dubai	8.25[a]	Transaction
2011	DIFC, Downtown Dubai	8	Transaction
2015	Standard Chartered Building, DIFC	6.25	Transaction
2016	The Edge, Dubai Internet City	6.75	Transaction
2017	Emaar Business Park, The Greens	7	Transaction
2018	U-Bora Tower, Business Bay	7.75	Transaction
2019	Emaar Business Park, The Greens	8	Transaction
2019	Emaar Square	c. 6–7[b]	Transaction

Source: Supplied by Andrew Love. Annotated notes from author research.
a This building was initially pre-sold with vacant possession (at a rate per sq ft by the master developer), explaining a higher NIY yield.
b Purchased with a 12-month unexpired term (no renewal), at a rate per sq ft. Approximate yield based on an assumed long lease at stabilized income, supplied by CBRE.

Table 5.4 Yield ranges across several commercial freehold property assets in Dubai (2018)

Asset type	Min (%)	Max (%)	Prime yield (%)
Office, long-lease, secure tenant	6.75	7.25	6.75
Residential building, long lease (accommodation)			
Office, medium lease, secure tenant	7.00	7.50	
Office, standard lease, multi-tenant	7.25	8.50	
Residential, annual, multi-let			
Retail, long lease, secure tenant	7.00	8.00	9.0 (all)
Retail malls	6.75	8.75	
Community malls	7.75	9.75	
Industrial			8.0
Healthcare and education			8.0
Hospitality			7.75

Sources: Data supplied by Andrew Love, modified by author. * If leasehold transactions these yields would be expected to be 100-200 bps higher, depending on the duration of the associated land lease.

malls is highest with regional and super-regional malls running at the lowest yield (6.75%) up to 9% for smaller retail malls and community malls (c. 9%). Prime hospitality although varied sits at 7.75%. While Table 5.4 provides an overview of yield evidence in Dubai over recent years, there has always been a challenge for valuers and advisors to secure more current yield data and often reference is made to transactions over a wide geography. Valuers typically categorize yields into prime and secondary districts for ease of comparison. Some assets, such as retail, are very rarely traded as these tend to be held in the ownership of the traditional retail groups within the region, wherein they are not only the owner of the underlying real estate but also the "sponsor" of the majority of the retail occupiers. One of the largest of these, Majid Al Futtaim (MAF) owns not only several of the main malls but also brands such as Carrefour which provide many of these the anchor occupancy.

In terms of purchasers of single-owned commercial buildings in Dubai, it appears many of the key benchmark transactions are those reported through local property funds and real estate investment trusts (REITs) (e.g., ENBD REIT; Emirates REIT; Izdihar Real Estate Fund and Amanat). Others remain as GCC-based private investors and HNWIs. International funds have spared little capital into Dubai's commercial property market. The apparent reasons for this lack of appetite for international institutional funds could be explained in Table 5.5 (see later).

Availability of data

Market maturity and transparency frameworks point to other reasons why Dubai's commercial property sector has lagged behind other major cities. A detailed analysis of this is provided in Chapter 8. Market maturity, however, is not static (Tiwari and White, 2010) and while market openness, availability/standardization of market information in Dubai was a historic issue, the tide is turning and more property data and information is being made available (see Chapter 8 for more detail). While commercial property market information is still lagging behind what is available in the residential market, more transactional information and wider economic forecasting can bring revived interest for Dubai's commercial investors.

Availability of suitable quality stock

The scope of commercial stock to invest in has been a historic deterrent to investment from both local and international funds (lack of suitable stock). Funds and financial institutions do not want the issue of fragmented build-ing ownership (in a strata law building) nor do they want the future cash flow uncertainty as you get with shorter leases. Regulation and lease law plays a significant role in the decision-making of institutions' asset alloca-tions. Table 5.5 highlights the comparison between commercial leases in Dubai and other western markets (using UK as an example). In the UK for instance, the majority of institutional investment is in London and the southeast of England, a by-product of where most institutions are based (Jones, 2021). Banking, finance, insurance firms have typically clustered in urban centres like this and this collective action has bred an appetite for investment/new development to accommodate blue-chip tenant; single-owned commercial investment. So what are the challenges presented for a similar institutional grade investment into Dubai, with the DIFC being a notably comparative financial district?

1   In London insufficient new office supply acts as a reassurance to insti-tutions that there is a formal control on tenant demand remaining in the buildings that they occupy. In Dubai, institutions may perceive that new supply is not so constrained and a new financial centre could emerge as the city evolves that competes with the specific asset.
2   Tenant risk is managed by regulation and lease law and institutions in London are more familiar with leasing practices in UK vs Dubai. It is much more common to see upward-only rent review clauses in a UK commercial lease than compared to a Dubai lease. The latter is more likely to be exposed to market volatility and reversion to open mar-ket rents. This lease profile presents more risk than an upward-only or stepped rent clause one might see internationally.
3   The process by which new real estate forms are accepted by financial institutions as suitable investments can be slow (Jones, 2021) and that might include their allocation to new markets like Dubai.

Dubai like the rest of the world has also seen increased occupational demand for industrial property (logistics, warehousing). However, it is not just offices

Table 5.5 Drivers of commercial property investment: UK vs Dubai

	UK	Dubai
Income	Leases are typically 5 or 10 years – sometimes even longer – a predictable, regular income, frequent upward only rent reviews.	Leases are typically 1–3 years, up to 10 (rare), market determined rent at lease expiry (more volatile as review periods are shorter and depend on lease length
Capital gain/growth	The value of commercial investment property is largely based on the reliability of the regular rental income v perceived risk. Planning laws restrict new supply/competition from entering market quickly	If the rental income is less predictable, there is a larger risk premium associated to the acquisition. Liberal development and "free-market" approach could lead to new supply shocks impacting value
Demand/ supply	Limited supply in prime locations means landlord offers less rental incentives/higher rental growth	Liberal planning/short development time means competition can be fierce and more incentives/lower rents need to be offered, outside core areas
Rent review legislation	If market rent is lower than current rent, the current rent continues. Fixed, stepped and inflation linked increases are also common	Upward in a rising market (limited to the rental index). In a falling market the rent reverts to market rent therefore income is less protected
Lease lengths	5–15 years, average 7 years	1–5 years, average 3 years. Longer leases have break clauses that can impact value (lower as less predictable)
Use of debt finance	Common to increase real returns.	Less common

Source: comparison adapted from IPF (2017) with Dubai commentary added by author.

that face challenges, industrial property/warehousing is also impeded by the lack of freehold ownership in Dubai's industrial zones. Many of the industrial locations in Dubai are currently on long-leasehold arrangements and there is uncertainty for institutional investors as to how this legal interest might leave them exposed to further legal risk. With even shorter development timescales (when compared to offices), logistics and industrial property and a fast-paced global evolution of industrial occupier requirements, alongside the land ownership structures, all contribute to a general lack of investment interest to financial institutions (or if interested, they would carry a high-risk premium). Prime industrial yields in Dubai have been sitting around 8% (+200 basis points to prime offices), highlighting the abovementioned point (when compared to other commercial yields – refer back to Table 5.4).

Institutional requirements and key drivers of commercial property investment

Let us now turn our attention to what institutions want and compare this against Dubai. This is to broadly highlight/explain why there has been a lack of financial institutions investing in Dubai. By evaluating these comparative practices, Dubai can work towards improving its ability to attract large sums of investment in its commercial property sector. Table 5.5 draws a comparison to the characteristics of the UK commercial market (that now attracts a significant amount of institutional money, circa 30% of UK commercial property is owned by overseas investors and 15% by UK institutions) versus Dubai, which as already stated has received a limited institutional interest.

Traditionally, financial institutions (including those who manage REITs) have invested in Grade A or "prime property". These characteristics can be defined under a range of building specifications and characteristics. Grade A office specifications typically include (as per the British Council of Offices (BCO)):

- Minimum ceiling heights of 2.6 m
- Ceiling void of 350 mm and floor voids of 150 mm
- Planning grid of 1.5 m by 1.5 m
- Passenger lifts (which have suitable speed and wait time parameters)
- Maximum opportunities for daylighting (based on 300–500 lux average, max depths for a shallow plan (15 m) or a deep plan (21 m) for natural light)
- Workplace rations of 80–100 sq ft per workstation

Institutions are typically risk-adverse and have specific investment parameters. Capital security (preservation) is more important than capital appreciation (or speculation). Income returns need to be stable and long-term with a lease mechanism that provides the owner with some hedge against future inflation. One might compare commercial lease clauses in Dubai to other mainstream investment markets and counter the argument that Dubai leases are not inflation-hedged. For financial institutions, the traditional strata management law (offices) and liberal property development regulations (new supply) as well as land title uncertainty (industrial) are examples of investment challenges in Dubai. That is not to say in the future institutional money will not enter Dubai, it looks likely only when product placement and more attractive leasing conditions are evident. A key challenge for foreign institutional capital is the direct competition it has with domestic and regional capital, which has typically driven down yields in Dubai to what is similar in a more established and mature market. As many global institutions still view Dubai as an emerging market, the risk-adjusted yield required at acquisition is often higher than what domestic and regional funds are willing to pay, thus providing a barrier to entry.

Once there has been a period of sustained letting and resale activity in Dubai then institutions may perceive these opportunities as credible as those in more mature markets. This acceptance leads to a fall in the associated risk premium and a rise in capital values. At this point, financial institutions widely incorporate the new real estate investment class into their portfolios. Long-track records of stable performance and market maturity will no doubt attract more interest. Perhaps institutions will evaluate the Dubai "risk premium" differential more favourably over time, as has been the case across different asset classes. Geographical arbitrage is opportune in Dubai if one is to evaluate the extreme presence of blue-chip, multinational tenants that occupy Dubai's commercial skyline. However, these step-changes in the institutional mindset and models, take decades to fruition. New risks, like ESG, are being presented within the traditional outlook of institutional investment and Dubai's relatively new, modern stock, alongside the right legal structures, may prove quite the offering in future years. "Brown discounting" is being overtaken by "green obsolescence" that will also impact where investors allocate their money in the coming decade. Chapter 11 provides a more detailed synopsis of these sustainability trends.

The next section evaluates the future outlook for commercial investment in Dubai, highlighting a range of new case examples that meet the institutional grades discussed above.

What is the future outlook for institutional investment in Dubai?

In the earlier parts of this chapter, reference was made to some of the barriers to institutional-grade commercial property in Dubai. The first barrier appeared to be related to the relatively short lease lengths. This is historically as a result of the short length of commercial licences issued for companies to trade, as these licenses have extended as have lease terms, especially within the free zones/special economic zones such as DIFC, TECOM and DMCC. This trend might provide an opportunity for longer lease propositions for institutional investors in Dubai. While Dubai's commercial property market represents a small proportion of global investment flows, new institutional-grade buildings are beginning to be developed in Dubai. The next section will explore a range of notable examples and highlights some of the latest institutional commercial buildings being developed in Dubai (see Box 5.2).

Another key driver of institutional investment into Dubai's commercial property sector may also come to fruition through the establishment of a UAE-specific financial yield curve. Box 5.3 elaborates on these developments and draws out a simple analysis as to how institutions can now more reliably price their property acquisitions through the introduction of UAE Treasury Bonds, as we see in other global markets.

Box 5.2 Institutional developments in Dubai

Case study: ICD Brookfield Place (new development)

When the joint venture between the Investment Corporation of Dubai (ICD) and Brookfield Properties was originally created, the aim was to create a fund that would acquire distressed assets or asset manage existing income-producing as a value-add proposition. However, having reviewed the potential opportunities in Dubai, the decision was made to develop something new (again reflected some of the challenges of identifying suitable Grade A institutional stock in Dubai at that time, supporting the above commentary). This brought about the new development of ICD Brookfield Place. The JV saw an opportunity in the ever-increasing demand for DIFC office space with tenant growth of 10%–20% per year. The development is really very specific to DIFC, where the vacancy rate is extremely low (<1%–5%).

The $1 billion development property was completed as a joint venture between ICD and Brookfield Properties, with the equity coming from the two joint venture partners, alongside bank financing (local as well as international). Key facts about ICD Brookfield Place are shown below:

Year completed	*2020*
Example tenants	Julius Baer, Natixis, Latham & Watkins, and Akin Gump
Typical lease length	5–9 years
Rent (ICD Brookfield Place)	AED250–350 psf + AED50 psf (service fee)
DIFC average rent	AED200 psf/upper limit AED340 psf (Savills)
Rent for flexible space (1,100–1,500 sq ft), 10 per floor. Average suite size 1,150sq ft (22)	AED650psf. JLL are currently leasing space from 1,000 sq ft up to 250,000 sq ft
Number of floors	54 (274th tallest tower globally), column-free corners, 925 ft height. The office tower and retail building will be connected by large pedestrian pathways with seven underground levels of car parking to accommodate 2,700 vehicles.
Office space	990,000 sq ft (each floor typically has 18,000 sq ft leasable space)
Retail space	160,000 sq ft
Sustainability rating	LEED Platinum/WELL certified
Other features	The development includes an 18,000 sq ft public area surrounded by restaurants and open spaces for regular arts and cultural events.
Joint leasing agents	CBRE and JLL

Case study: Banking on Dubai: Institutional commercial real estate in Dubai (build to suit)

These case studies are examples of corporate occupiers seeking institutional-grade offices in Dubai. With the lack of stock available, both organizations opted for a build-to-suit solution.

- **HSBC Tower,** $250 million, 20-storey building in Emaar Square, Downtown Dubai. LEED Gold, 58,500 sq ft (available for leasing to external tenants on the sixth, seventh and eighth floors). The fifth floor and floors 9–20 were occupied by HSBC. The occupational density has been set at 1:97 sq ft. Ground floor gymnasium for staff/occupiers. Completed in 2018. Typical floor plans across the GFA of 400,000 sq. ft. highlight potential configurations under mixed (150 sq ft per desk); open plan (97 sq ft per desk) and cellular layouts (258 sq ft per desk).
- **Standard Chartered Tower**, Downtown Dubai. The building is located next to the HSBC Tower. Eighteen-storey building. Office units start at 50 sq ft. Like with HSBC Tower, the building began as the Head Office for Standard Chartered (levels 6–13). Levels 1–5 are shell and core and the ground floor is a retail branch for SCB. The total gross floor area is 23,500 sq m. (253,000 sq ft). LEED Gold (Shell and Core) and LEED Platinum (for SCB fitted offices). The occupational density has been set at 1:90 sq ft. Completed in 2012.

In both these instances the international banks were creating an "on-shore" HQs, outside of the regulations of the DIFC. With the example of Standard Chartered Bank, the bank has two Dubai HQs (one onshore and one offshore). Future changes to the regulatory environment such as dual-licensing could potentially mean in the future these could be housed in a single property albeit with internal demarcation.

Another similar corporate occupier example has been the new VISA HQ opened in 2021.

- VISA (CEMEA), Dubai Internet City: A G+5 floor office building (including hotdesking, breakout areas and collaborative spaces). The building has been designed based on the operational requirements of VISA. The property was developed by Sweid and Sweid and accommodates 500 staff over 100,000 sq ft

Box 5.3 Will the introduction of Treasury Bonds bring more institutional investment?

In April 2022, the UAE Government announced the launch of Treasury Bonds (first issued in May 2022 in two/five-year bonds, with a view of later bringing in ten years). It is the first time that the UAE has done so. This will help property professionals build the UAE dirham-denominated yield curve from which to benchmark real estate pricing often taken from this country-specific risk-free rate of return (applying a ten-year T-bill). In October 2021, the UAE Government has begun the launch of USD-dominated T-bonds. Prior to this some price benchmarking had been done via EIBOR rates.

Analysts expect there to be a reduction in the yield premium over time. These now appear to be trading at a 2% yield (https://cbonds.com/country/UAE-bond/).

So how does this affect commercial real estate pricing?

Globally, treasury bonds have been at historic lows (US, UK, Australia). Real estate professionals habitually connect low-interest rates and bond yields to low property yields and high prices. Academics internationally have always been interested in the relationship between the T-bond (ten years) often assumed to represent the country's risk-free rate of return and property yields. These studies have sought to find a long-term "yield gap" between T-bonds and real estate. The outcome of this would then provide pricing information from which analysts and funds could determine whether real estate was under or over-priced. A UK study by MSCI in 2018 showed a 20-year long-term average of 3.5% yield premium for commercial real estate. A similar US study by PMA in 2018 found the US prime office yield premium to be 2%. Research also found that whilst the future increase in bond yields will probably impact property yields, they are unlikely to increase in parallel, provided rental growth is increasing at the same time (IPF, 2014). A range of global studies indicates that the real estate premium above ten-year T-bonds to be 2%–3.5%. There has been a tendency for the risk premium to decline over time. In addition, in periods of property exuberance, as was with the run-up to the GFC in 2007, yield premiums were negative or inversed for real estate. This meant investors were accepting lower yields on real estate than they were on ten-year T-bonds, indicating continued expectations of future rental growth and capital gains. An appreciation of the property risk premium long-term allows some rational pricing in property markets, so what might the introduction of Treasury bonds in UAE mean for property investors and the transparency of initial yields for commercial property in particular?

What is the correct yield for Dubai commercial property?

Based on an application of a simple pricing equation ($Y = RF + RP - G + D$), a suitable property yield can be derived from a consideration of the following benchmark data:

- Risk-free rate (RF) is 3.0%* (US treasury + 1%)
- Assume a risk premium (RP) for Dubai property to be 4% (+200 bps above US average)
- Assume rental growth (G) of 3% (split between real growth + inflation linked)
- Assume depreciation (D) of 1%
- Initial Property Yield (Y) = 3+4-3+1 = 5% (grade A institutional quality)

*In June 2022, the three-year T-bond (AED) was fixed at 3.24%. A higher rate is anticipated for the ten-year bond later this year.

Therefore, simply speaking if prime yields are currently 6%–7% then offices in Dubai look relatively cheap/attractive.

Regardless of the simplicity of the above analysis (and its means of demonstrating a rational pricing hypothetically), the imminent introduction of UAE ten-year treasuries will allow for institutions to undertake formal benchmarking of real estate pricing via the yield curve. This global pricing strategy will likely attract new interest in measuring property prices and risk in Dubai and potential future acquisitions.

Indirect real estate investing in Dubai

Given the scale and nature of global commercial real estate assets, there has been some appeal to the growth and development of indirect real estate funds. The notable driver for investing in REITs in other global markets is their tax advantage/efficiency. While Dubai is a tax-free jurisdiction and this seemingly would hold no additional advantage for Dubai investors, the diversification benefits of REITs still hold. The pooling of investor capital means a more diversified exposure to the commercial property market (Baum and Hartzell, 2012), which offers some benefits away from direct real estate investment. The challenge, like with international institutions, is the lack of suitable assets for capital deployment. REITs typically hold a structure whereby their allocation to different income-producing assets per sector is capped. Therefore, there are stock selection challenges for local REITs. The next section looks at the options for indirect property investing in Dubai.

What options do investors have to gain exposure to commercial property markets in Dubai?

The two main alternatives would be via listed stocks on the DFM, such as Emaar Malls (this would give an investor exposure to the retail assets operated by Emaar, for example, flagship assets like Dubai Mall). Alternatively, there is a limited number of REITs available in Dubai, listed on the DFM/Nasdaq Dubai. Both options rely on an assessment of asset quality either within the listed firm (e.g., Emaar Malls) or the listed REIT (see case examples below). The tax advantage of a REIT seen in other markets is perhaps not present in the Dubai market and it would therefore be more apparent to assess the performance based on the asset quality in the REIT versus that of a listed operational equity option.

REITs are funds that invest in real estate. The DFM REITs platform offers greater diversification for investors to efficiently access rental income streams from wide-ranging, underlying real estate assets without buying them or spending large amounts of capital. The DFM REITs platform enables issuers to raise capital from their property without selling it, through its income stream.

Two REITs are currently listed in Dubai, include: Emirates REIT, whose IPO in April 2014 was the first REIT listing in the GCC, and ENBD REIT, which was listed in March 2017 (as cited on Nasdaq Dubai). Both REITs are Shari'a compliant. The structuring of Shari'a compliant REITs is similar to conventional REITs except for the Shari'a guidelines and principles. To be Shari'a compliant often means excluding from the portfolio, for instance, some of the best-rated tenants in the banking sector. Conventional banks charge their clients "interest" and seek "profit"; therefore, this would prevent them or the buildings they occupy from being included in the REIT (as with other non-Shari'a compliant examples). This puts a limitation on the choice the REIT manager has when selecting commercial assets.

Key characteristics of listed REITs in Dubai

In the DIFC, in addition to the REIT being closed-ended, it must also (as cited on Nasdaq Dubai):

- Be publicly listed on a regulated exchange
- Invest no more than 30% of the funds in properties under development
- Distribute 80% of its annual net income to shareholders (versus 90% in the US)
- Borrowing must not exceed 70% of the NAV of the fund

The fact that the majority of the net incomes are distributed to the share-holders then it resembles the income profile of direct commercial real estate investing. Furthermore, the income is not based on a reliance of a single commercial property, yet the combined income across a selection of assets invested in by the REIT asset manager. Below is a highlighted summary of the case examples of Emirates REIT (case study 1) and Emirates NBD REIT (case study 2).

The performance of Emirates REIT since its inception is shown below (Table 5.6)

Before listing on Nasdaq Dubai in 2017, ENBD REIT's predecessor, Emirates Real Estate Fund, delivered cumulative returns of 32.11% and consistently distributed dividends. Table 5.7 highlights the recent price performance since listing.

From 2017 to Q1 2022, the fund lost 44% of its Net Asset Value (from US$297 million to US$167 million). As seen, the historic performance of the Dubai REITs has not always been positive, and, in the case of Emirates NBD, they have recently discussed taking the REIT private due to liquidity issues (The National, 2019). However, the REIT today remains listed and instead focuses on repositioning its portfolio towards "alternative" assets in sectors such as logistics (warehousing) and healthcare that offer longer lease agreements and more stable income streams (The National, 2020).

Case study 1 Emirates REIT

Emirates REIT was founded in 2010, but listed on Nasdaq Dubai in DIFC in 2014. As of March 2022, the REIT currently has 11 properties under its active management (USD758 million) as listed below:

Properties under management	Asset class
Index Tower, DIFC	Office, Retail and Car Parking
Office Park, Dubai Internet City	Office, with GF Retail
Loft Offices, Dubai Media City	Office
Building 24, Dubai Internet City	Office with GF Retail
Indigo 7, SZR	Office with GF Retail
European Business Centre, DIP	Office with Retail
Trident Grand Mall, Dubai Marina	Community Mall
Lycee Francais Jean Mermoz, Al Quoz	School
GEMS World Academy, Al Barsha	School
Jebel Ali School, Akoya, Dubailand	School
School in DIP	School

Source: Emirates REIT.

Case study 2 Emirates NBD REIT

Properties under management	Asset class	Gross yield (%)	Occupancy (%)
Souq Extra, Dubai Silicon Oasis	Retail Centre	10	99
The Edge, Dubai Internet City	Office	8.8	97
South View School, Dubailand	School	6.5	100
Uninest, Dubailand	Student Accommodation	11.6	100
Al Tharaya Tower, Dubai Media City	Office	5.7	43
Burj Daman, DIFC	Office	7	69
Arabian Onyx Tower, Barsha Heights	Residential	10	97
Binghetti Terraces, Dubai Silicon Oasis	Residential	8.9	68
DHCC 49, Dubai Healthcare City	Office	9.1	78
DHCC 49, Dubai Healthcare City	Office	8.1	67
Remraam Residential, Dubailand	Residential	6.5	46

Source: ENBD, as of 31/03/2021. Portfolio | Dubai Property Investment Company | ENBD REIT UAE.

Table 5.6 Historical share performance of Emirates REIT since inception (2011–2020)

2011 (%)	2012 (%)	2013 (%)	2014 (%)	2015 (%)	2016 (%)	2017 (%)	2018 (%)	2019 (%)	2020 (%)
+5.3	+10.9	+25	+10.3	+14.2	+10.2	+10.6	+4.3	−5.1	−20.7

Source: Emirates REIT.

Table 5.7 Historical performance ENBD REIT (2017–2022)

Period	ENBD REIT	Return on reinvested dividends	Total return
1 Year	−4.44	0.00	−4.44
3 Years	−7.25	0.00	−20.22
5 Years	−14.38	2.58	−53.99

Source: ENBD, annualized returns over 1, 3 and 5 years.

Was it a step in the REIT direction for Dubai?

The development of REITs in Dubai was a positive move in terms of improving its path to maturity by providing investors a diverse range of investment products. However, the funds have not been immune to the challenges of both market cycles and finding suitable investment products in the market. In mature markets, the investment performance of REITs is tied to long-term income, whereas Dubai typically provides short-term commercial leases (less income security). Therefore, it is likely to suffer a more volatile return profile (like equities) when compared to a more secure long-term investment product seen in international markets. In addition, there are some further complications in asset allocation as typically these funds are Shari'a compliant and this goes down to the types of tenants occupying the commercial property when looking at potential acquisitions. Critics have also called into question whether REITs offer any additional tax advantage for investors, which is typically a key driver in other international markets.

REITs in Dubai are still relatively limited and their performance has been mixed, perhaps with many still preferring to invest in direct residential property in Dubai (as the earlier statistics suggested). However, as institutional stock enters the market, opportunities do exist for more development in REITs and other indirect investment products. The appeal of well-managed large-scale shopping malls, offices and hotel properties would certainly be appealing to a range of global investors.

Conclusions

The region has a potential for larger growth of inward capital from international fund managers, but a lack of information to make informed decisions and liquidity risk had impeded a more widespread uptake. The analysis of the commercial property market in Dubai has indicated that, despite early challenges, it is evolving and evidence points to more institutional money entering the market. While the region as a whole has historically seen low levels of funds entering, evidence in the latter parts of this chapter has indicated that there is an appetite to acquire the right commercial property in Dubai. Foreign capital has had to compete with local/GCC funds with the latter pricing assets at lower capitalization rates. Since the liberalization of foreign ownership, ownership structures and expensive finance has been a barrier to entry, as well as not being able to source the "right" institutional grade stock. In the last few years, international asset management firms and developers have proved that there is a commercial appetite for single-owned commercial assets, be it as a build-to-rent or build-to-suit proposition. The case examples provided are indications that more funds will be deploying capital in Dubai if the right buildings are developed, something we are likely to see more of in the next 5–10 years.

A key area of growth in purchasing Dubai real estate has been the supportive legislation for both purchase and sales and leasing. The next chapter moves on these discussions to provide a comprehensive overview of how real estate is traded in Dubai.

References

Baum, A., and Hartzell, D. (2012) *Global Property Investment: Strategies, Structures, Decisions*, Chichester: Wiley-Blackwell.
British Council of Offices. BCO specification for offices. https://www.bco.org.uk/.
Data Finder (2022). https://www.datafinder.ae/ (Accessed June 2021).
DLD (2021) Commercial Property Price Index Q3 2021. https://dubailand.gov.ae/en/open-data/research/commercial-property-price-index-q3-2021/.
Emirates REIT. https://reit.ae/.
ENBD. http://www.enbdreit.com/reit/portfolio/#funds.
Hoesli, M., and MacGregor, B. (2000) *Property Investment: Principles and Practice of Portfolio Management*, London: Routledge.
IPF (2014) Implications for Property Yields of Rising Bond Yields (June 2014). https://www.ipf.org.uk/static/uploaded/a5e70eed-38a4-44ee-9cb896efe356dd1b.pdf
IPF (2017) 'Understanding UK Commercial Property Investments: A Guide for Financial Advisers' https://www.ipf.org.uk/resourceLibrary/understanding-uk-commercial-property-investments-a-guide-for-financial-advisers--december-2017-.html
JLL (2022) The UAE Real Estate Market - A Year in Review 2021. https://www.jll-mena.com/en/trends-and-insights/research/the-uae-real-estate-market-a-year-in-review-2021.
Jones, C. (2021) *Urban Economy: Real Estate Economics and Public Policy*, London: Routledge.
Mirza, A. (2009) Down and Out, The Economist. https://www.economist.com/finance-and-economics/2009/04/23/down-and-out.
The National (2019) ENBD REIT to Delist from Nasdaq Dubai. https://www.thenationalnews.com/business/property/enbd-reit-to-delist-from-nasdaq-dubai-1.948512.
The National (2020) ENBD REIT to Reposition Its Portfolio by Focusing on 'Alternative' Assets. https://www.thenationalnews.com/business/property/enbd-reit-to-reposition-its-portfolio-by-focusing-on-alternative-assets-1.1012263.
Tiwari, P., and White, M. (2010) *International Real Estate Economics*, London: Palgrave Macmillan.

Part C

Professional practices

6 Sale, purchase and leasing practices in Dubai

This chapter will begin by outlining the real estate sales process in Dubai, bringing attention to the main methods used by agents in the purchase and sale of property. An overview of the buying process is then explained, high-lighting the typical range of transaction fees for the sale and disposition of various property transactions in Dubai as well as distinguishing the different forms of property interests that can be purchased and by whom. The rest of the chapter will largely focus on leasing practices in Dubai as it becomes a critical part of reference to professional practice later in this book. Real estate valuation, investment and development all require a basic understanding of leasing documents, as these create the legal terms that derive the cash flow earned by the owner and/or investor. When buying property, an investor would want to know how is the future rental income decided. How easy is it for the rent to be increased? Who is responsible for the operation and maintenance of the property during the period of occupancy? An essential requirement of real estate practice is therefore understanding laws and regulations. In most global real estate markets, a system of land and property tenure prevails, which broadly speaking offers the following features:

- A right to purchase, use and lease property;
- Legal rights of sufficient duration to allow investment;
- Real estate assets are traded and can be used for capital raising (through secured lending);

Therefore, real estate, or strictly speaking, rights in real estate have become tradable assets due to the above characteristics. Each of these parameters typically need to exist to have a functional real estate market. Governance and regulation control these rights and these laws and regulations in turn control how real estate is perceived in a particular jurisdiction. For instance, planning control and a planning framework provides certainty to real estate investors and developers about new supply. The specific property interests that can be purchased (freehold/leasehold) also govern how attractive a market is.

The first section examines the purchase and sale of real estate in Dubai.

DOI: 10.1201/9781003186908-9

Estate agency in Dubai: Law and practice

This section is designed to introduce several key areas of estate agency practice which outline the key legal aspects a practitioner or prospective buyer should be aware of. It is important that you have a sound knowledge of the current legal background to estate agency, marketing methods and techniques and an understanding of clients' objectives, when acting as an agent or dealing with property transactions in Dubai more generally. This chapter is not designed to be a fully comprehensive overview of every legal aspect of property, instead it is aimed to provide the core knowledge required for a property practitioner or graduate surveyor seeking a holistic overview to major sales and leasing practices in Dubai.

Transacting in property does call for special skills and this is acknowledged under bespoke legislation, such as that displayed in Dubai practice (predominantly via RERA) that requires brokers to be licensed. As with other emerging economies, Dubai had recognised the need to control or regulate estate agency practices, particularly given the importance to investor confidence to the economy. In Dubai, Real Estate Regulatory Agency (RERA) has established practice certification for estate agency brokers. Much of this qualification is in line with international aspirations of raising professional conduct, focusing on core areas of legal practices, such as code of ethics, property registration and transfer procedures, standard documentation and introduction to the relevant governing bodies or authorities. We see further evidence of similar schemes operating in other international markets. Box 6.1 highlights the current local laws and regulations relevant to the estate agency profession in Dubai.

Box 6.1 Current laws and regulations in Dubai real estate agency

Bylaw No. (85) of 2006 Regulating the Real Property Brokers states the full range of laws and articles related to agency practices. Those commonly referred to in a standard transaction would include:

- No Person may conduct the Brokerage activity in the Emirate unless he is licensed by a Competent Entity and is registered on the Register (Article 3)
- A Broker registered on the Register must apply annually to the Division for renewal of his registration (Article 13)
- Registered Brokers must comply with professional code of conduct in accordance with the Code of Ethics (Article 14)
- Brokerage agreement must be in writing and must state the names of the contracting parties, the specifications of the Real Property, and the Brokerage terms (Article 26)
- Two percent is the typical agency fee for standard residential transactions (though can be negotiated) as stated in Article 27
- Subject to Article (30), if a party enters into separate agreements with several Brokers for the same Brokerage or negotiation assignment,

and only one Broker succeeds in concluding the transaction, that Broker shall be exclusively entitled to the full fees.

• Currently (as of 2022), a seller can appoint up to 3 property agents to list their property.

When looking to offer property in the market place, an agent and the instructing client will need to consider a range of factors which will lead to decision as to the best possible method of disposal. There is nothing however in legislation to prevent an agent from initially applying one course of sales and then changing to another, so long as a contract has not been entered into between buyer and seller. In broad terms, there are three main methods of sale:

• Private treaty
• Tender
• Auction

When deciding upon which method of sale to recommend a client to use, consider such factors as the client's objectives, public accountability, current and future market conditions and the anticipated demand for the property plus disposal timings. Each method will bring with it its own style of agency practice in relation to the terms of engagement, marketing (sales particulars, advertising) and which professionals are required to take an active role in the sale of the property. A summary of each of these three main methods of sale are discussed below:

Private treaty

Sales by private treaty are undoubtedly the most popular form of sales disposal (estimated to be 90%–95% of all sales). In such an arrangement, prospective buyers are free to negotiate with the seller, without commitment and in the open market. The process starts with an agent, once having cleared terms of engagement and due diligence, offering the property in the open market (or restricted market if so agreed by the client). When going through the process of marketing by private treaty, an agent will need to consider the following stages:

• Agreeing terms of engagement with seller (Form A)
• Carrying out due diligence enquiries
• Seek any legal information
• Inspect the property
• Agreeing with a marketing strategy
• Offering the property to the market
• Carrying out viewings
• Agreeing terms of engagement with buyer (Form B)

- Checking pre-approval finance of buyer (if buying with mortgage)
- Passing buyer to mortgage broker (if required)
- Agreeing on a purchase price in principle with buyer and seller
- Passing to solicitor/legal representative (if required)
- Agreeing major heads of terms with buyer and seller
- Both parties signing agreed heads of terms (as an MoU or Form F, securing a buyer's deposit, normally 10% of the purchase price (held by the agency as a cheque))
- Clearing of seller's mortgage on property (typically drawn from buyer's mortgage, within 30 days of title transfer). For a cash seller, this step is removed
- Clearance of NOC (seller's service fees cleared and confirmation of no outstanding debt from seller)
- Title deed issuance at DLD representative office
- Issuing fee invoice (as per MoU)

Further details of the buying process are discussed in the next section ("How to buy a property in Dubai?").

Tender

This method of marketing is used when there is a good level of interest in the property. It can be used when marketing a property by private treaty, when several prospective purchasers display interest in the property, especially where offers have been made in excess of a guide price. Sale by tender is occasionally used as the preferred method of sale for commercial property or development land, most notably in a buoyant market. By doing so, the interested parties are invited to submit their best and final offers by a closing date. A tender is therefore a marketing method used in order to exercise control over the marketing process. The process begins with the estate agent inviting all interested parties to submit their 'best and final' offer in writing by a defined closing date. For commercial transactions in Dubai this accounts for approximately 5-10% of transactions, and is rarely used for residential sales.

Auction

Auctions are commonly used for the sale and disposal of specialist property. There is also a need for auctions for the disposal of properties in Dubai that go into foreclosure. Emirates Auction is the most common portal where properties from lending institutions can dispose of assets (via Dubai Courts). The clearest advantage of an auction is that a sale can be achieved in a relatively short timeframe (especially as the financial institution is looking to liquidate the defaulted loan/property as soon as possible). Buyers interested to purchase distressed properties can do so via the online portal, subject to paying a 20% deposit (depending on the value amount mentioned on the auction pages) paid as a manager's cheque (secured as cleared funds). This is required to participate in the auction. The remaining balance of the purchase amount must be paid within nine calendar days after the auction date. There are specific requirements of buyers who register via the portal including the depositing of a refundable security deposit (as referenced above). If the buyer wants to increase the bid amount, then the 20% security deposit also needs to increase. More details on the processes involved are found at: https://www.emiratesauction.com/en/propertydxb/onlineauction.aspx

The next section explains some details in terms of the buying and selling process in Dubai. It is not overly complicated nor too different from other global markets; yet many prospective buyers are unfamiliar with how the process works in Dubai.

How to buy a property in Dubai?

A property owner is under no obligation to use to services of an agent. However, it is widely regarded that the use of an agent will increase the likelihood of a sale, as the property is receiving wide exposure in terms of advertising and marketing. If a vendor was looking to take on the services of an estate agent then they would be required to sign a contract (RERA Form A), outlining the terms of service for agent and vendor (for instance, % of sales commission; whether the agreement is exclusive or not; methods of sales and marketing). After Form A has been signed, it will be approved by the DLD's Trakheesi system, which will assign the property advertisement a permit number. This system was introduced to ensure that property listings can be verified and reduce the number of "fake" listings that historically plagued the listing portals.

The use of the internet and more specifically property portals and listing apps has ensured a wide range of potential purchasers are introduced to various properties, with virtual touring facilities offering an alternative to conventional face-to-face meetings. Below are a number of leading real estate search portals when looking to buy or sell in Dubai:

* Property Finder (www.propertyfinder.ae)
* Bayut (www.bayut.com)
* Houza (www.houza.com)

More recently, these portals have introduced more value-added services, including information relating to recent sales transactions (as per the DLD registry); service fee charges; an anticipated gross or net yield. This move towards improving the data transparency for prospective purchases allows buyers to make more informed decisions about the specific property.

As with vendors, a prospective purchaser is likely to have some interaction with an agent in order to at least establish a good idea of the current market value and to seek advice and property details. In that sense, agents offer expertise and advice not only to the vendor by estimating a property's worth and marketing it but also to the purchaser as a means of exposure to a suitable property. If a prospective buyer wants to work with an agent on buying a property in Dubai then RERA Form B would be signed.

If the purchaser wants to buy they make an offer through the agent and then it is the agent's responsibility to promptly inform the vendor of this offer. Estate agents must treat all prospective purchasers fairly and must not hold back offers for any reason. Hence, another role of the agent is to establish the "most suitable" offer in the best interests of the vendor and it is not always the highest offer that should be accepted. More often than not, a purchaser/vendor who can proceed quickly is more attractive (that would typically relate to whether either one or both parties were holding an outstanding mortgage debt or planned to take on a new mortgage in the case of the prospective purchaser). Once an offer is accepted, the agent has a responsibility to confirm the offer in writing, but still has a legal duty to pass on any further offers. The offer is only binding once both parties (buyer and seller) have signed an MOU (often referred to by RERA as Form F). During these stages, agents will hold where necessary, any deposits paid to secure the purchase of the property, typically 10% of the purchase price. In some cases, the buyer may also ask the seller to put down 10% deposit in the same manner avoiding the likelihood of gazumping or gazundering throughout the sales process. One should appreciate that normally the agent's main interest lies with the seller whom he tries to obtain the best deal. A buyer tends only to be of principal concern to the agent when the purchaser is paying a fee to an agent to find a property.

The main function of an agent is to arrange the sale of property. In addition, they advise on the price, negotiate between purchaser and vendor and manage advertising, as well as conduct surveys, offer mortgages and referrals to other financial advisors. More recently, agents have also started to liaise on behalf of clients with third parties and offer other services, for instance financial products or home inspections. Even in the latter stages of a property transaction, agents can be vital and a fundamental requirement is that they progress the sale smoothly through to a successful conclusion. To this end, agents may require an additional fee for such service and it is not uncommon for agents to charge an additional AED5,000 for managing the sales process. Once the buyer and seller have agreed on a price and signed the MOU, the property would be ready for the official transfer at the DLD. Nowadays, a property transfer is typically undertaken at a local transfer office. The timeframe for transferring the property from the seller to the buyer is dependent on a range of factors, notably those relating to the financial position of both parties. If the property seller has no outstanding debt/mortgage and the buyer is purchasing in cash, the transfer can be done within seven days of signing the MOU. If one of the parties/both parties are mortgaged, then the transfer time takes a little longer as there are the

Table 6.1 Typical range of transaction fees for the sale and disposition of various property transactions in Dubai

Type of property sale (or transfer)	Fee (% of value or stated amount)
Registering a sales contract	4% (or 4.25% for Islamic Ijara finance, with the additional 0.25% relating to rent-to-own contract)
Registering a warehouse sales contract	AED10 psm (not less than AED10,000)
Registering a mortgage contract (and/or transferring a mortgage via refinance)	0.25% of the loan amount
Discharging a mortgage	AED1,000
Registering a long-term lease contract	4%
Registering a Usufruct right	2%
Issuing a title deed	AED250
No objection certificate (NOC) from master developer	Circa AED500–1,500 (depending upon the master developer), paid by seller
Estate Agents Fees	2% (negotiable between parties), often expected to be paid by buyer

associated due diligences required including an independent property valuation (RICS firm via lending bank); clearance of the mortgage between Bank A (seller) and Bank B (buyer). In both instances, most property purchases require the seller to get a NOC (no objection certificate) from the master developer to ensure the property is free of any dues (typically bank finance and service charge payments). Table 6.1 highlights the typical range of transaction fees that are part of the property purchase process in Dubai.

The form of agency and purchase and sales process has been outlined. The next most suitable discussion would be to examine the forms of ownerships in Dubai. As with other global markets, owners and occupiers of real estate in Dubai can either own a freehold (perpetual) title or a leasehold (restricted) title. The next section will start by examining the legal parameters of freehold and leasehold ownership in Dubai.

Freehold vs leasehold interests

Freehold is a common global term that typically refers to an absolute ownership. Most freehold interests in Dubai are also seen to be the same and legally follow elements, such as:

* *Possession:* the exclusive right to the occupation and use
* *Duration:* long-term ownership
* *Scope*: inclusion of specific land and property

Freehold rights in Dubai are legally evidenced by a title deed registration at the DLD. This type of tenure retains all perpetual rights to use or dispose of the land as the owner wishes, subject to certain legal restrictions. This form of ownership is permitted to both UAE/GCC nationals and foreign nationals and is allowed within specific areas in Dubai known as freehold communities/or districts (see Table 6.2).

However, there are some further distinctions between freehold areas. *Restricted freehold* (or "private freehold") can be bought only by UAE/GCC nationals. Some prominent restricted freehold areas include Mirdiff; Al Barsha; Bur Dubai and Deira. Dubai Investment Park (DIP) is an example of a commercial area of restrictive freehold. UAE/GCC nationals can own anywhere in Dubai (freehold and restricted freehold areas), whereas foreign nationals are only permitted to own in designated freehold areas, of which there are currently 30. When acquiring property, it is common for the "transfer" to take place at transfer offices, conveniently located across many areas in Dubai. Once the freehold interest is purchased, the legal owner can then either occupy the property or lease it to another party. Common practice in leasehold ownership is for the entire land and property to return to the freeholder at the end of the lease (tenancy), albeit there are provisions

Table 6.2 Forms of property ownership in Dubai

International term	Dubai term	Relevance to practice
Freehold	Absolute ownership subject to common law and legislation. UAE/GCC nationals are allowed to own any real estate interest (absolute, usufruct and musataha rights) anywhere in Dubai Foreign nationals (and expats) are permitted to own in designated "freehold" areas (currently 30).	Perpetual unless stated on the title deed. Transfers at the DLD DIFC is the only freehold area to have its property laws based on English Common Law.
Leasehold	Usufruct (for up to 99 years) Musataha (for up to 50 years, renewable)	All leases have to be registered via Ejari/RERA. Under UAE federal law if a tenant continues to occupy the premises after the lease expiry without objection from the landlord then an automatic renewal of the same period prevails. Modification of the lease terms requires a 90-day notice prior to the lease expiry. A tenant can assign or sublease with landlord consent

Source: Author's own

imposed by law that protect the tenant from unreasonable actions upon renewal or vacation of the premises. Income return from an investment property is borne out through the presence of leasehold ownership.

In Dubai freehold areas, the management of the apartment buildings is typically run similarly to that of a commonhold interest in other markets. This "commonhold" legal structure allows for freehold ownership of individual flats/units in a designated area. The rest of the building or master community is owned and managed jointly, through an owner's association. Therefore, when units are sold to individuals, a framework of rights and obligations then exists between the owners of each flat and the owner's association. Briefly, this framework would consist of the following points:

- A unit may be a flat (residential) or an office or shop (commercial). The DLD will create a registered title per unit (and reference any specific parking spaces included)
- The common areas, for example, the lift and the stairs; entrance lobbies; building amenities (pool and gym, for instance), the car park, all of which will be managed by an inhouse or third-party service provider appointed by the Owner's Association.

Annual management and maintenance costs are then charged as a service fee to each unit holder based on a proportion of the ownership in the building (the sq ft stated on the DLD title deed defines the area from which AED per sq ft service fees are apportioned). RERA regulate the account of these annual service charges to ensure owners are proportionally and reasonably charged for the management of the common areas. These charges are typically collected on a per-quarter basis, with annual adjustments at year-start and year-end.

Leasehold similarly relates to exclusive possession of land held for a fixed term (of years or even months) in consideration of rents payable to the owner of the freehold property/or owner of a leasehold interest (if subleasing is permitted). A leasehold interest differs from freehold ownership as it relates to "a right to exclusive possession" which is for a limited number of years. Leasehold ownership, like most international markets, relates to those property interests permitting local and foreign nationals to occupy a property for a specific number of years (ranging up to 99 years). In Dubai, these leasehold interests are further separated into *Usufruct* (up to 99 years) and *Musataha* (50 years, renewable). A range of additional transfer rights exist with the presence of gifted and granted land ownership to UAE nationals from the Ruler of Dubai. Table 6.2 (shown above) provided an overview of the various definitions of freehold and leasehold practice in Dubai.

The discussions will now examine leasing practices in Dubai in some further detail. These laws and regulations have an impact on a range of professional practice, including asset management, valuation and investment appraisal. In addition, characteristics of a lease differ significantly, and these would ultimately lead to a different cash-flow characteristic for the investor.

It would be important when mapping out a DCF appraisal or valuation to take specific note of the individual terms that impact the future cash-flow assumptions of the asset. Whilst buildings can appear similar, a commercial office in DIFC with poor lease terms and poor-quality tenants would be worth significantly less than a similar building in DIFC under single ownership on a long lease with a strong tenant. In this next section, a range of fundamental laws and practices will be outlined to provide a student or new practitioner a comprehensive overview of how real estate is legislated in Dubai. It would be important to link the content of this chapter with other later chapters where these legal or leasing practices can be applied.

Leases in Dubai

This section is designed to provide an overview of the leasing practices in Dubai. The observations made are summative and do not cover every legal aspect of leasing in Dubai. Instead, the discussions are focused on highlighting key elements of leasing that will be of interest to a real estate practitioner. I will also draw your attention to how leasing practices in Dubai compare to other international markets and thus highlight the implications this might have for real estate practices, such as valuation and investment appraisals. Understanding leases is also critical in investment and asset management decision-making, providing the analysis a cash-flow roadmap.

Leases also drive responsibilities for both the landlord and tenant throughout the occupancy or holding period and thus are critical in understanding how the cash flow is likely to look. This is further exacerbated in Dubai as often leases are of a short-term duration, so it also gets us to think more strategically about the appropriateness of how we value property in Dubai, when we typically do not see commercial leases beyond ten years. These lease structures also have implications for attracting institutional capital to Dubai, after all many institutions will have long-term liabilities, for instance, within a pension fund.

In my opinion, the lease and occupation trends are the most critical aspects to understand when evaluating a real estate acquisition or property valuation. After all, without income security, we would see inherent risk built into these assets. Conversely, a commercial asset that has a long lease arrangement to a single multinational tenant, like Microsoft or Google, has much stronger underpinnings. In these cases, Dubai would offer some geographical arbitrage when considering risk, a risk reduction in part provided by the quality of the occupier and also the key institutional lease terms. If I reflect on my home market in the UK, we would expect to see a five-year upward-only rent increase at review, a cash-flow characteristic very different to what we would expect to see in Dubai. According to CORE (2020), the typical lease length for commercial space under 1,500 sq ft is 1–3 years with an option to renew (for an equal term) than the original lease term. Space in excess of 1,500 sq ft, 3–5 year terms can be more common. Residential leases are typically annual leases (renewable). Table 6.3 draws a comparison of the

Table 6.3 Comparison of common lease terms in Dubai vs global markets

Lease term	Dubai (UAE)	UK	US	Australia/NZ
Term	Short; 3–5 years	Long; 10–15 years	Negotiable, generally any length	Moderate; 3–10 years
Breaks	Available for longer leases	Options to break	Negotiable	Options rare
Renewals	Automatic if tenant holds over and landlord does not object	Negotiable. Law gives tenant right to renew, but this right can be waived	Negotiable	Common
Rent basis	Net	Net. Most leases fully repairing and insuring	Gross	Net or gross
Free rent	1–6 months/variable	Wide range. Between 6 and 33 months on a 10-year lease.	Negotiable; 1–12 months are typical	Negotiable
Escalation	Local laws limit increases	Negotiable. Typically, every 5 years. Sometimes to market, usually upward only.	Negotiable	Every 1–2 years
Security	1–3 months	Negotiable	Negotiable	3–12 months
Fit-out	Sometimes included in the rent	Tenant pays	Landlord	Tenant pays
Landlord's broker	N/A			
Tenant's broker	Tenant pays (5%–10%)	Tenant pays	Landlord	Tenant pays
Right to sublet	Negotiable with landlord and regulatory approval	Allowed with restrictions	Common	Common with approval of landlord
Transparency	Limited	High	High	High (AUS) Limited (NZ)
Space measurement	No single standard. *IPMS Offices	NIA	Rentable area, usable area, net usable area	NLA (Property Council of Australia) and Net BOMA (New Zealand) *see PCA
Building classification	Grade A, B, C	Grade A, B, C	Class A, B, C	Grade A–D

Source: Waters (2019).

Data extracted from CBRE Global Office Report (2015)* IPMS (International Property Measurement Standards) will be adopted in Dubai as mandatory*Building grade classification is set out by the Property Council of Australia in their Guide to Office Buildings Quality and is determined by factors such as NLA, floor plate size, environmental, mechanical, tenant riser, lifts.

major common lease terms in Dubai with those in several other mature international markets, highlighting key differences in leasing practices.

In Dubai, there is the presence of a unified tenancy contract, that covers industrial; commercial and residential properties. A unified tenancy contract (lease) registered via Ejari is governed by three main laws, and include:

- Law 26 of 2007, that governs the landlord and the tenant;
- Law 33 of 2008, that includes some amendments to the abovementioned law; and
- Law 43 of 2013, that governs rental increases within an existing or expiring lease agreement

These laws will be explored in relation to the context of the discussions below. A unified tenancy contract does permit additional terms that can be agreed between the parties, however these cannot conflict with the laws above. These were introduced largely to give more protection to the tenants especially in the residential market, in relation to evictions and rental increases (see details later).

For leases in Dubai to be protected by statutory regulation then they must be registered (on Ejari). The next section outlines the process of lease registration in Dubai.

Lease registration in Dubai

Leases in Dubai are somewhat standardized (unified) and their legality is governed through their registration on the Ejari (a term that means "my rent" in Arabic) system, either as a business entity (commercial) or a private individual (residential). A unified Ejari tenancy contract can be downloaded from https://ejari.dubailand.gov.ae/PublicPages/ejari.htm. The system and some of the legal aspects of lease registration and management via the Ejari system are identical for both commercial and residential leases. The purpose of lease registration on the Ejari system in Dubai is to ensure the rights and responsibilities of both the landlord and tenant are kept within a central system. External processes, such as connection of utilities; and visa sponsorship are bound to the process of registering relevant leases via Ejari. It also ensures if a dispute between parties arises, then the first port of call would be RERA, the regulatory authority that oversees the Ejari system and manages disputes within it. Only cases involving registered leases will be heard through the Rental Dispute Committee (RDC). More details about the typical disputes that may arise in Dubai are discussed later. Table 6.4 provides information on what documents need to be included when registering either a residential or commercial lease in Dubai.

The registration of leases once signed is typically executed and paid for by the tenant. Many have considered this a simple and beneficial system set-up that authenticates a rental contract, especially now since much

Table 6.4 Required documents to register a lease via Ejari

What documents are required to register a lease via Ejari?
Passport/residence visa
Emirates ID
Tenancy contract (the "lease")
Copy of title deed of the leased property (owner/landlord)
DEWA connection deposit receipt (new contracts)/Recent DEWA bill (renewal)
Owner's passport
For commercial leases, a copy of the trade licence is required

---

encouragement has been placed upon doing this via the government's digital/online platforms. This is an annual process or follows the renewal period stipulated within the lease.

What constitutes a lease in Dubai?

In its simplest definition, a lease is a legally binding contract between a Landlord (Owner) and a Tenant (Occupier). Article 3 of Law 33 of 2008 extends the use of leases to both land and properties, with Article 4 governing the relationship between landlord and tenant via the lease. Such leases must have certainty as to:

- the parties to the lease (stating the names of Landlord(s) and Tenant(s));
- the property being leased (building name; location; property size; type) and property number (and DEWA premise number for linked utility services);
- the purpose of the tenancy (permitted use and property type);
- the period of the lease (length with beginning and end dates);
- the amount of rental to be paid for the lease (in words and numbers including the method and frequency of payments);
- be registered on Ejari

Whilst the parameters of a lease are fairly standardized across both residential and commercial leases, for instance, rental escalation is currently bound by the same RERA rental index law, there are some variations that need highlighting between the two. The use of non-standard leases does exist in the commercial market in particular where the landlord is seeking the adoption of fixed or % based rental increases. Tenancy Laws (Law 26 of 2007, as amended by Law 33 of 2008) offer parties clear guidance on what is required for a lease to be valid and enforceable and stipulate any rights to dispute committees for all Ejari-registered leases.

As mentioned, the nuisances within the lease are critically important in providing both income and capital security to the owner. That said, every lease is different. Commercial terms do differ from those seen in a residential lease and vice versa. A real estate practitioner; investor or commercial valuer will ultimately be much more considered around the lease terms that impact their cash flow or rental escalation assumptions. Therefore, it is important to discuss several market norms when it comes to commercial leases, that perhaps vary much more than the residential counterpart, the latter is typically an annual lease commitment to an individual. There are a range of lease agreements that commercial occupiers can enter in Dubai. These include four main types of commercial leases (Dubai Chamber (2012); Bayut (2018)):

- *Gross lease:* rent paid by the tenant on a fully serviced arrangement, where the landlord is responsible for other expenses like maintenance and insurance (similar to the UK IRI leases). It is typical to see a gross lease on a single and multi-let office buildings.
- *Triple net lease:* a base rent paid by the tenant plus an additional payment relating to a number of other property costs, such as insurance and repairs (similar to the UK Full Repairing and Insuring (FRI) lease). The base rent + FRI costs puts the burden of the property maintenance onto the tenant, thus this type of commercial lease has become very popular for landlords in Dubai
- *Modified net lease:* Expenses are shared between landlord and tenant, commonly used in commercial leases for a range of industrial properties, retail or multi-tenanted properties.
- *Land lease:* The tenant pays for a long lease on a parcel of land, and constructs the building to occupy. At lease expiry, landlord takes possession of land and building

Now we know the importance of leases and their ability to be different to one another, let us look at the main Heads of Terms that make up leases in Dubai, many of which we also see in other international markets (see Box 6.2).

Box 6.2 Leasing in Dubai – Headline terms

a   **Rent:** This can either be fixed (based on market conditions, expressed as an annual amount in the lease, and paid in advance (quarterly; bi-annually and in some cases annually) or turnover (based on business performance, more common in retail properties, provided as a base rent with a top up rent based on turnover).

b   **Lease length:** Residential leases are typically one-year/annual tenancies with a right to renewal. Commercial leases range from 1–5 years, with some institutional leases extending up to ten years.

c   **Renewal rights:** Automatic renewal if neither party has notified the other before 90 days of the expiry of the lease. Amended terms to the lease must also be given before 90 days of expiration date.

---

d   **Break rights:** Residential leases typically bind for the full 12-month period but do come with a 2-month notice penalty (landlord's discretion). Longer-term commercial leases carry a break clause (supported with a 3–6 month notice period/penalty).

e   **Rent deposits:** Residential leases have a one-month security deposit. Commercial leases typically range from 1–3 month rent as a deposit. Rental deposits can be used for a number of tenant issues including unpaid rent; damage to the property that is intentional or via gross misuse; lost keys or passes. Wear and tear provisions apply.

f   **Rent reviews:** Both residential and commercial rental increases are linked to the RERA rental calculator, whereby rents can be increased if the current passing rent is a certain % below market rents (as per the index). As per *Decree No. 43 of 2013*, rental increases are bound by caps and this applies to all properties in Dubai, only after the first year of a lease. A commercial lease may include a rental increase for the lease term, often appearing as fixed increases.

g   **Service charges, insurance costs and other outgoings:** Rent payments typically include service charges (residential) but can be additional for commercial tenants (see Chapter 7).

h   **Assigning, subletting (alienation):** Assignation or sub-leasing is at the full discretion of the landlord (Article 24 and when permitted Article 30 to be followed).

i   **Repairs:** The repair obligations sit with the landlord (major). Repairs up to AED500 (residential) and AED2,000 (commercial) can be the responsibility of the tenant.

Landlord and tenant obligations

Leases in Dubai carry a number of standard provisions for both the landlord and tenant. Table 6.5 provides a basic overview of examples pertaining to both the number of landlord and tenant obligations.

As we see in other international markets, leases in Dubai also carry provisions for termination clauses. Provisions within leases allow the landlord to end their relationship with problematic tenants if there are any breaches of the lease (via Article 25), for example; late payment of rent; failure of the tenant to use the property for the end use expected; unapproved sub-letting; or breaches in any of the lease terms (30-day notice required). In commercial leases, the landlord is also permitted to evict the tenant if the tenant fails to occupy the property without legal cause for 30 days (continuous) or 90 days (non-continuous) in any one-year period (unless otherwise agreed).

Table 6.5 Standard landlord and tenant obligations

Examples of typical landlord obligations	Examples of typical tenant obligations
No alterations to the premise after the lease is signed. Otherwise, landlord will be liable for any damage caused	Pay the rent when due as stated on the lease (Article 19)
Provide property in good condition that allows the tenant to immediately occupy and use the property, unless otherwise agreed (Article 15, Law 26 of 2007)	Pay the security deposit to cover the condition of the premise at the end of the lease (excluding wear and tear), Article 20
Landlord is responsible for maintaining and repairing any defect or damage that may affect the Tenant's intended use of the property, unless otherwise agreed (Article 16, Law 26 of 2007)	No alterations and renovations without Landlord's consent (Article 19)
Landlord must provide to the tenant all rights to quiet enjoyment, including access to the building's common services and utilities. If the landlord disconnects services the tenant has the right to file a case with Police and RDC for damages. Article 34	Return the property to the Landlord in good condition, with only reasonable wear and tear (Article 21)
	Pay all government fees and utilities, unless otherwise agreed (Article 22) Not to remove tenant improvements unless otherwise agreed (Article 23)

Source: Extracted/modified from standard Dubai lease, referencing Law 33 of 2008 and where stated Law 26 of 2007; DLD (2020).

Another example of time-bound legal notices includes the parameters to which for other legal reasons (not a tenant breach) the landlord can evict the tenant. These all require the serving of a 12-month notice of eviction (via Notary Public or Registered Mail)) and include:

- If the landlord wishes to sell the property;
- If the landlord wishes to use the property for his own personal use or direct relative (landlord has the burden of proof that he has no other properties for use);
- If the property requires demolition or reconstruction;
- If the property requires major renovation that cannot be completed with the tenant in situ (a technical report required).

Following on from one of these permitted legal reasons, there are a number of stipulations that would need to be adhered to, including:

- If the eviction is granted for the landlord's personal use, the property cannot be leased for the following two years (three years for commercial property).

- The law states that a 12-month notice must be served at lease expiration (except in the case of one-year leases).

As a final point on maintaining a landlord relationship when a property is sold or transferred to another party, Article 28 states that the current tenant will remain in accordance to the tenancy contract signed by the seller. Therefore when purchasing a property in Dubai with a sitting tenant, the buyer would need to follow Article 25 and serve a 12-month to gain occupancy for the same legal parameters listed above.

Assignment and sub-leasing

A lease in Dubai can be assigned (or sub-leased) but this needs the expressed permission of the landlord before doing so. Further practicalities from a corporate real estate perspective are discussed in Chapter 7.

Further to these main headline terms (see Box 6.2), it is common for both residential and commercial leases to clearly stipulate the notice periods needed to abide by both parties. For example, it is typical for a tenant or landlord to provide a 90-day notice to the other party if they wish to change or agree to new terms prior to the expiry of the lease. If both parties are silent then the lease would renew automatically on the same terms as agreed in the current lease for an identical lease duration. Where a rental increase is sort after by the landlord, but they failed to exercise notice before 90 days to the tenant then the expiration of the 90-day notice period would also mean that right to a rent increase is lost.

Managing rental increases

As mentioned above, *Decree No. 43 of 2013* enforces rental increases. They are bound by caps as follows. If the tenant was paying:

- Ten percent or less than the average similar rent, the landlord cannot increase the rent (no increase permitted)
- 11%–20% less than the average similar rent, the landlord can increase the rent by no more than 5% of the current rent
- 21%–30% less than the average similar rent, the landlord can increase the rent by no more than 10% of the current rent
- 31%–40% less than the average similar rent, the landlord can increase the rent by no more than 15% of the current rent
- Over 40% less than the average similar rent, the landlord can increase the rent by no more than 20% of the current rent

A draft law (discussed since 2019) has proposed a three-year fixed rent for residential tenancies in Dubai and if passed, there would be no permitted rental increases during the first three years of signing the lease. This would bring further transparency on the residential cash flow for both a landlord and investor.

Managing rental disputes

Decree No. 26 of 2013, relating to rental disputes, attempts to settle all claims within a 75-day period. The DIFC follow a different legal recourse. The dispute process is relatively straightforward and comes with the benefit of all leases that have been registered via the Ejari system. The party who wishes to file a dispute must open a case with the RDC which involves paying a fee, based on 3.5% of the annual rent amount (AED500 minimum/AED20,000 maximum). The process is then heard by a maximum of two interrelated parties: Arbitration and Reconciliation (15 days); Department of First Instance (30 days). If the case is further disputed the case would be filed with the Department of Appeal (30 days), although these are restricted to cases over AED100,000 (with other conditions applying) based on the payment of a 15% fee.

Commercial leases are more complex than residential leases and require careful management. The most prevalent aspects to the context of this book would be those related to operational risks through additional costs. Scott et. al. (2017) highlights that a commercial tenant should understand whether the lease permits the landlord to charge a "service charge" for additional costs of services (and insurance and other landlord costs). When it comes to either an investment appraisal or a market valuation, a valuer must also consider the specific lease terms in order to match the cashflow assumptions, such as net operating income (annual rent; lease length or rental escalation), as well as who is responsible for the repairs, insurance and services within the property.

The practical implications of this are discussed further in Chapter 7 and 9.

Conclusions

The development of real estate legislation in Dubai was pronounced between 2002 and 2010. Following the creation of Dubai's RERA in 2007, the market has seen it playing an increasingly public and crucial role in developing and supervising Dubai's real estate regulatory framework. These legislative developments were initially for financial institutions related to registration of lender's pre-mortgage interests and were extended to cover loan-to-value lending criteria; landlord and tenant responsibilities as well as the enhancement of valuation practices.

Buying property in Dubai is relatively straightforward and the process is clear and precise. The chapter has provided a summary of these key stages and how property can be bought and sold in Dubai. Leasing regulations have improved significantly over the last 10–15 years, much-balancing regulation in the favour of tenants to ensure they are treated fairly and processes are transparent, most notably in managing rental increases. This chapter has provided a clear summary of the latest legislation, however, a clarification from legal professionals is recommended as laws and legislations can significantly vary overtime.

Leases in Dubai are governed by a series of legislation, namely, in relation to the security of tenure and rental increases. Currently, both commercial and residential leases are governed by the same legislation. Commercial lease agreements are usually in the range of 3–5 years (certainly less than ten years) and are considered tenant-friendly in relation to automatic rights to lease renewal and clear guidance on how landlords are able to vacate their premises. Rent for lease extension or renewal is negotiable and a rental disputes committee is available if parties cannot agree. The RERA rental index also governs clear guidelines on how much the rent can be increased by relevance to the location and current passing rent. If parties fail to follow these guidelines, again, a dispute committee is available to ensure compliance.

A key part of understanding leases in any market is to be in a better position to evaluate income security. A cash flow can be anticipated through leases and legal rights to income. When these laws and processes are clear and transparent, there is less risk attached to the process.

A comparative analysis of common lease terms has shown the differences between standard Dubai leases versus "institutional" leases for UK commercial property. As shown, Dubai's commercial markets have been characterized by much shorter leases, frequently with annual changes in income and often with non-recoverable costs; costs which would vary over time. These differences have important implications for real estate valuation methods and techniques in Dubai. New valuers or those coming from other marketplaces must start to examine the specific risks set out in the lease rather than opting to "hide" assumptions under the all-risk yield (ARY) approach. Despite the ARY being suitable in a transparent market, with a high number of comparable evidence, its suitability historically in Dubai has been called into question. It is still applied in Dubai, even though the availability of data is somewhat lacking and impaired by the segregation of freehold and non-freehold transactional evidence. Since the introduction of the valuation registration schemes in August 2016, future development towards international best practice in Dubai has looked more likely. The forthcoming chapters examines the evolution of property data in Dubai (Chapter 8) which also further supports the discussion of valuation practice in Dubai (Chapter 9). A key point to remember when working and investing in an international market is what appears to be the same may not be the same, terms identically in a written form can have different legal meanings.

The next chapter extends discussions on commercial leasing practices by offering a closer look at corporate real estate management in Dubai.

References

Bayut (2018) https://www.bayut.com/mybayut/renting-commercial-properties-dubai/.

CORE (2020) https://www.core-me.com/commercial-guide.

DLD (2020) 'Tenancy Guide' https://dubailand.gov.ae/en/about-dubai-land-department/tenancy-guide/#/

Dubai Chamber of Commerce and Industry (2012) Signing a Commercial Lease, *Business Information Series*, 1 (3), pp. 4–5.

Scott, J., Arnott, I., and Balfe, A. (2017) The Fundamental of Commercial Leasing in Dubai, Al Tamimi and Co. https://www.tamimi.com/law-update-articles/the-fundamentals-of-commercial-leasing-in-dubai/

Waters, M. (2019) A critical examination of property valuation variance in Dubai, PhD thesis.

7 Managing a global commercial real estate portfolio in Dubai

Chapter 1 highlighted the economic and geographic benefits of the UAE, occupying a strategic location between Asia, Europe and Africa. This has attracted many businesses to set up and operate from Dubai as a preferred base in the Middle East. Dubai offers firms access to new business as well as provides staff a range of "lifestyle" benefits. The Global Competitiveness of Dubai has been measured frequently across different benchmarking studies. One such example is the JLL Global Cities research that measures global competitiveness. The metrics within the report include:

- Corporate presence: A critical mass of corporate HQs in key professional areas such as banking, finance and media
- Gateway functions: A two-way flow of people, investment, trade, tourism and information
- Scale and market size: A scale of population and market to support agglomeration and provide diverse opportunities for firms, people and capital
- Infrastructure: An efficient connectivity of firms, workers and visitors
- Talent: A diverse pool of skilled labour
- Specialization and innovation: An ecosystem to produce commercial specialisms; knowledge and game-changing innovation
- Soft Power: A global brand and identity which is attractive and projects the city's core values

In 2020, Dubai was highlighted as an attractive proposition for international businesses. The key drivers for office occupancy in Dubai centre around the ease of doing business. According to Trading Economics, the UAE ranked 16th out of 190 countries globally for the ease of doing business, a measure on how the regulatory environment is conducive to business operations. Favourable corporation taxation and labour cost and availability also appear important drivers. But how do businesses who set up in Dubai evaluate their property holdings? What options are available to international corporate tenants?

DOI: 10.1201/9781003186908-10

Within the corporate sector, property is typically held for operational reasons (above a return on capital through investment that we might see internationally). As discussed in Chapter 5, a limited number of foreign institutions are buying their commercial space. Operational risk and market risk are perhaps reasons that explain a much higher leasing propensity in Dubai over other jurisdictions. Another has simply been the limited supply of institutional-grade office space in Dubai. There are several historic examples whereby if the space is not in the market the corporate occupier has gone about building its own (perhaps the most referenced to date are the Standard Chartered and HSBC buildings). This trend of build to suit is becoming more apparent in future commercial development. However, not all businesses can buy their real estate in the locations they operate within.

Before setting up the parameters of analysis for Dubai, there are certain distinctive aspects relating to real estate that distinguish it from other factors of production for a business globally including:

• Real estate has several measures such as cost, value, utility, branding and corporate governance that form where and how a business locates.
• Real estate is complex, illiquid and long term, therefore given these asset characteristics, it may be misaligned with an organisation's short-term business strategy.

Corporate real estate management focuses on how businesses use space and the way in which efficiencies are measured has been well debated (see Weatherhead, 1997). Property wastage and mismanagement promoted the evaluation of space requirements. Much of the international literature on corporate real estate management has centred on the cost savings that are apparent in the operational side of real estate asset management. Property is the second largest cost to a business (after its people/salaries), therefore, business performance is maximized when firms and occupiers are conscious of improving efficiencies and reducing "wastage". Global businesses setting up international operations need to evaluate the decision based on comparative information, most rudimental in the analysis being both space and cost. However, a key challenge today is that businesses are located in different international markets and need to evaluate those same "currencies" to their home markets, yet a lack of consistency in the way in which property-related information is recorded and made available presents an efficiency barrier in any corporate decision-making.

The most compelling example of this recently has been the measurement of space, which up until 2014, utilized a range of different measurement standards (BoFM, RICS, SISV, AI). A JLL study in 2012 found property space measured to a different professional standard could differ by as much as 24%, making it challenging for corporate occupancy costs to be fairly and consistently measured internationally. IPMS now provides a business a "fair" analysis of the space requirements globally. Dubai was one of the first governments to adopt IPMS back in 2014, prior to that no common standards of measurement were used (IPMSC, 2013). Space requirements nowadays can be quite straightforward with the international institutional grade building classifications and IPMS well understood. There is still a consistency issue with lease cost information. The information presented in this chapter looks to provide a suitable summary of key differences in the way lease costs are built up in Dubai.

As this book is focused on Dubai, the next sections will provide a summary of key corporate decision-making for businesses located in Dubai, framing this around a hypothetical case example.

The core analyses include:

- Where to locate and how much space to take?
- What are the typical leasing costs for corporate occupiers in Dubai?
- Should a business own or lease their corporate space?

How do firms in Dubai decide where to locate and how much space they need?

A corporate entity seeking a presence in Dubai has two main options (Deloitte, 2021):

1 Establish a presence in the UAE mainland, i.e., outside the Dubai free zone areas. Previously, businesses located in mainland locations would need to partner with a UAE national shareholder (51%) or a company wholly owned by UAE nationals. However, in 2020 Decree Law No. 26 now permits 100% foreign-owned companies. If a firm took up space in a non-freehold area in Dubai it would be restricted to a leasing model as foreign property ownership is not permitted to non-GCC. Furthermore, the business would be required to take a minimum office space of 200 sq ft (as required by the Dubai Economic Department to issue a new trade licence).
2 Establish a presence in one of many of Dubai's free zones. The selection of an appropriate free zone is most likely related to the type of business sector or operation the firm is seeking. Dubai's free zone is typically linked to a particular sector, e.g., healthcare; education; media; financial services. This option would enable a corporate to either own or lease a business premise.

In Dubai, there are some specific regulations that will govern what and how much commercial space is needed (legislated as a minimum amount of office space per worker, for instance). Commercial space and its demand are tied to the size of the corporate workforce. Property space is required for corporate licencing in UAE as a further example, tied into work visas for staff. This has typically been in the region of 100 sq ft of office space per employee (refer to case study).

How do leasing costs for corporate occupiers in Dubai compare versus international markets?

According to the CBRE 2019 Global Occupier Survey, the average office occupancy cost globally is $67.10 psf, versus Dubai average rates of $85 psf. As a comparison, Hong Kong ($322), London ($222), New York ($196) and Beijing ($182) were noted as the most expensive global locations for office occupiers. As with most markets, there are spatial differences in office rents across Dubai (refer back to Figure 5.4 in Chapter 5).

A typical commercial leasing transaction for an office or industrial property in Dubai would be 2–5 years in duration, up to ten years with the inclusion of suitable break clauses (Colliers, 2019). When analyzing occupancy costs in Dubai there are some additional parameters that are relevant to highlight, both in terms of office space leasing practices as well as lease fundamentals. Fit-out costs and rent-free periods are both "cyclical incentives" dependent upon the prevailing market conditions. A very specific component to leasing costs in freehold areas vs onshore areas is the presence of a tenant's service charge, that might be an additional AED20-30 psf to the agreed headline rent (Colliers, 2019). A typical commercial lease in Dubai will state the amount representing the tenant's proportion of the service charges, yet there are differences across Dubai with many buildings in onshore areas offering rents inclusive of a service charge.

When it comes to managing leasing costs during a current lease in Dubai, Chapter 5 and 6 highlighted the legal requirements of renegotiation notice periods (90-days prior to lease expiry); rental escalation being either fixed or based on the rental index; subleasing requires the written consent of the landlord; and the potential of early termination penalty clauses. Additional considerations that impact a business and their leasing cost might include workspace ratios which historically are significantly larger in Dubai, in the range of 100-150 sq.ft. vs 80-100 sq.ft in UK. (Colliers, 2019).

As noted above, there are some structural leasing practices in Dubai that do increase the cost of leasing for commercial occupiers, compared to other international markets. For instance, typically a tenant is paying rent + an annual service charge that adds to the total cost of occupancy (above and beyond the headline rent).

The above commentary has provided a broad indication of a commercial occupier's core considerations when taking on space in Dubai. To put this discussion in some context, the final section provides a case study analysis of an international corporate occupier with leased office premises both in Dubai and the UK. The intention of this case study is to examine the key criteria that a foreign business might consider when looking at its commercial occupancy costs and efficiencies. By reviewing the following case study, it is hoped that readers will be able to draw out the practicalities of the earlier discussions within this chapter (and Chapter 6) and understand better the context of commercial leasing practices in Dubai (Box 7.1).

Box 7.1 Case study: Managing an international corporate portfolio in Dubai and the UK

This case study explores the likely range of considerations a corporate occupier would place if they were to have properties in Dubai and another international location. The firm has a UK headquarter based in a prominent city centre (6,500 sq ft) and in January 2022 acquired a Dubai-based firm offering solutions to UAE businesses having seen a large increase in tech-based advisory within PropTech and FinTech. After the acquisition of the firm, the business holds three leased properties in its portfolio (one in UK and two in Dubai), the details of which are shown below:

	UK HQ	*Dubai 1*	*Dubai 2*
Location	CBD, city centre	Dubai Media City	Dubai Silicon Oasis
Net Internal Area (sq ft)	6,500 (Class A)	12,280 (Class A)	4,470 (Fully-fitted, Class B office)
Lease Start Date	January 2017	March 2017	August 2020
Lease Term	10 years with 5-year rent review	5 years	2-year lease
Total rent per annum	£123,500	AED1,960,000 (base rent of AED1,653,000 plus tenant to pay AED25psf service charge)	AED421,000 (inclusive)
Sub-leasing	Negotiated in lease	No subleasing without landlord consent	Strictly no subleasing permitted
No. of Staff	86	80	34
Workstations	66	80	48

The main aim of the business is to improve space and cost efficiency.

Case study questions

1   By applying suitable metrics, how different are the costs and occupancy efficiencies in Dubai versus the UK HQ?
2   What other considerations (beyond real estate) would be important to consider when evaluating this portfolio?
3   List the specific considerations in the Dubai portfolio that you would address in a corporate property strategy.

If we were to evaluate the current portfolio, we need to examine this information and data more carefully. The common metrics used in cost and space utilisation assessments are based on cost (standardized to size (per sq.ft.); per employee and per workstation) and area (standardized to per employee and per workstation). Under the assumption the headline information in the table above is correct and has been verified (e.g. property measurements) then an initial data analysis can be undertaken. Table 7.1 provides a summary for reference.

Table 7.1 Initial data analysis

	UK HQ	*Dubai 1*	*Dubai 2*
Location	CBD, city centre	Dubai Media City	Dubai Silicon Oasis
Net internal area (sq ft)	6,500	12,280	4,470 (Fully-fitted)
Annual cost psf	£19.00	£31.92[a]	£18.84[a]
Total employees	86	80	34
Total workstations	66	80	48
Annual cost/employee	£1,436	£4,900[a]	£2,476[a]
Cost/workstation	£1,871	£4,900[a]	£1,754[a]
Area/employee	75.58	153.50	131.47
Area/workstation	98.48	153.50	93.13
Employee/ workstation	1.30	1	0.71

a   Assumed AED to £ at £1=5 AED.

From the initial analysis, a business might discuss the following points:

• How do the costs per sq ft compare across the three premises? For instance, Dubai 1 is significantly more expensive psf than the other two in the portfolio. Is this a result of a cyclical trend on the lease or due to location? How critical is this location to the business and can the efficiency of the occupancy be improved?

- How attractive are the passing rents in the three offices versus the current market rent? (i.e., Is the business underpaying or overpaying for the space?).
- Going to the cheaper area in Dubai (DSO) – What implications does that raise for the business? (customers; clients; staff retention; workforce lifestyle from surrounding amenities). Office location will play a big role in attracting and keeping staff (transport, ease of getting to work, surrounding amenities and places to live). What action is imminent as per the leases? (renewal, 90-day notice periods, rent reviews/renegotiations, sub-letting).
- How is space being utilized currently in the portfolio? Currently, Dubai 1 has a significantly larger sq ft per workstation compared to UK HQ and Dubai 2. Despite the observed notes stating Dubai typically has higher workspace ratios, an analysis could run a scenario based on 100 sq ft workstation model to "stress test" the impact of how the business is using the space.
- UK HQ appears to be using some form of flexible working as there are more employees than workstations (1.30 ratio). Can Dubai operate a similar model? If so, what impact would that have on the amount of office space required? There is a legal requirement in Dubai that a corporate must have c. 100 sq ft per work visa (i.e., Employee). How does this regulation impact the initial analysis above? Based on this, what opportunities exist for offering staff flexible work practices?
- Both Dubai properties have leases expiring soon. What options are available? Is the cost of Dubai 1 likely to increase or decrease (see point 1)? Is a central location critical? If the business shifted to a cheaper location what impact might this have on the workforce/client base? Is it practical to keep the two offices in Dubai for different business services?
- Ultimately, the core business strategy is to improve the space/cost efficiency in the current portfolio. A range of factors could be drawn out in the analysis having a balanced view and looking at things from different perspectives and crediting the likely impact such decisions would have on both real estate and staffing issues.

Other legal considerations for the Dubai portfolio might include:

- Corporates can be restricted to the industry-specific regulations of a free zone (finance, media, healthcare, logistics). There are generic free zones that cater for wider business services. Subsequently, businesses need to locate in the correct free zone otherwise banking services and corporate bank accounts cannot be operational/opened.

- As per the laws and regulations of the free zones and civil laws of the UAE, a company registered with any free zone cannot conduct their business or activities outside the specific free zone, at a different free zone or inside UAE mainland. There are exceptions for a free zone company to trade or operate outside of their free zone and to use a mainland distributor or mainland third-party provider.
- There are minimum amounts of office space required per employee (refer back to legal need of 100 sq ft per employee). Therefore, with Dubai staffing at 114, the firm would have to take a minimum of 11,400 sq ft of office space. Currently, it has 16,750 sq ft, so the organization has the capacity to employ a further 53 staff (an increase of 50%). Another consideration here is that would the firm make any real cost savings if it introduced work-from-home practices. On the above assumptions of 100 sq ft per employee, the firm only has the capacity to reduce office space by 30% (down to 11,400 sq.ft). If it was to go more aggressive it would need to lose staff and reduce the overall operational staff in Dubai.

One key area of corporate decision-making will be what to do with the existing tenancies. Both Dubai properties are close to their lease expiry/renewal (March 2022 (Dubai 1) and August 2022 (Dubai 2)). Box 7.2 details the rent review/renewal process for commercial property in Dubai making reference to the RERA rental calculator. All Ejari-registered leases in Dubai are bound by this rental index.

Box 7.2 Hypothetical rent review on Dubai portfolio

Using the RERA Rental Index Calculator (shown at the link below), advise on the new rents and other considerations that comply with the leasing practices in Dubai. Advise the UK firm on to the disposition of its two existing properties in Dubai. Make the assumption that both leases expire at the month's end (i.e., 31st of the month).

https://dubailand.gov.ae/en/eservices/rental-index/rental-index/#/

When inputting the current lease information into the rental calculator, you should be able to draw the following advice.

- Dubai property 1: Media City: lease expires end of March 2022. The current rent is AED160 psf. When renewing real property lease contracts, there will be no permitted increase in the rental value as the premise is already above the average rent of similar units in Media City. The rent for an office in Dubai Media City is in the range of 107–145 (AED/sq ft). If both parties were silent on the renewal, it would continue on the same terms. Assuming that 90 days' notice was not served, then an automatic renewal would occur. Under UAE Federal law, if a tenant continues to occupy the premise after the expiry of the lease term without objection from the landlord, then the lease shall be deemed renewed for a similar period on the same terms. The renewal period is restricted to a similar period or for one year whichever is less. Therefore, in this case, a new lease would automatically renew and expire on 31st March 2023.

- Dubai property 2: Dubai Silicon Oasis: lease expires August 2022. The current rent is AED94 psf. The rent for an office in Dubai Silicon Oasis is in the range of 47–63 (AED/Sq ft). There is no increase in the rental value. If the firm wished to vacate the premise they may need to provide a 90-day notice to the landlord (i.e., 30 May 2022). The specific notice period is now contained within a Dubai lease.

Another major consideration for a corporate would be the decision of whether to buy or lease its premises. The final part of this chapter will examine this debate in more detail. The rise of property as an investment asset has risen globally in the last 50–60 years as more businesses have taken up leasing of commercial property. This has called into question how businesses hold space during their occupancy with owning versus leasing being the two distinct options. A change in business practices; leasing practices and global accounting standards have added further complexity to the decision-making. However, Golan (1999) cites flexibility, control and financial merit as being the primary considerations for most businesses. Global occupiers are faced with two key decision points when it comes to owning or leasing commercial real estate:

- Does the organization want to use the property as an asset, capital appreciation or for future borrowing or equity release (via sale and leaseback) or is this capital investment best employed in the business rather than real estate?
- How important is flexibility vs potential short payback periods that might be evident during a particular market cycle?

In Dubai, international corporates have tended to opt for leasing over buying (with some notable exceptions explored in Chapter 5). As a new market, businesses perhaps also favoured taking on a small amount of office space to road-test the commercial profitability of its core business. One notable risk in owning is the uncertainty of the capital value at any particular point in time. Chapters 2 and 5 pointed to the volatility of prices and higher risk premiums on a potential acquisition, especially in the early years, that may have hindered businesses owning, a risk that most global businesses setting up operations in Dubai may have not been willing to take. As the market matured and continues to do so, these market risks may subside. As shown in Chapter 5, large multinationals and banks are appearing to be now considering owning space in Dubai. This may breed in the future a growth in institutional sale and leaseback deals in Dubai, as we have commonly seen in other more mature global markets, offering a structured exit of disposing of their freehold interest. However, as discussed above, the firm must be located in a freehold area in order to be able to purchase its commercial premise.

The next section examines the decision in the context of the Dubai portfolio.

Own vs lease in Dubai?

The decision on whether to own or to lease is largely driven by a change in corporate attitudes. In the past, many organizations enjoyed the flexibility that leasing offered new entrants to Dubai. The business uncertainty and unfamiliar rules and regulations that were granted since 2002 for foreign property ownership meant many businesses relied on a leasing model. The scale of operations when setting up in Dubai was often small and about testing the waters to their presence in a new market. Taking on a small amount of leased space initially seemed a preferred approach to corporate decision-making. That is not to say businesses are not expanding and taking up more space in Dubai and we certainly observe more evidence of international firms occupying their own premises and gaining more control over that asset/building through owning. While in certain parts of the economic and business cycle, flexibility is a preferred option, it also exposes the occupier to more volatility and mid- to long-term uncertainty about what those property leasing costs might be. While long-term leasing is an option for businesses, transactional costs for leases beyond ten years are equivalent to the 4% DLD purchase costs. Buildings in prime locations will do well and demand for occupying their own premises is certainly on the rise in Dubai. The main advantages and disadvantages of freehold ownership and leasehold in Dubai are summarized in Table 7.2.

A number of international studies have evaluated the own vs lease decision via a DCF financial analysis. For example, Haynes and Nunnington (2010) compared the NPVs of both options in the UK context. Box 7.3 provides a buy versus lease scenario relating back to the earlier case study example, the assumptions matched to the specifics of the Dubai commercial market.

Table 7.2 Freehold (buy) vs leasehold (rent) decision-making in Dubai

Freehold – Advantages	*Freehold – Disadvantages*
Allows the business to design/build specific occupier requirements, against the historical challenge of finding top-quality office space.	Volatile, opaque market historically may mean that the opportunity cost of owning is eroded
Build to suit model has allowed occupiers a future source of equity when sold on a leaseback agreement	Capital gains may be inversely impacted by depreciation rates and/or other unanticipated real estate risks, e.g., geo-political
May serve as a suitable hedge against inflation/provide investment returns	Owning commercial real estate can be management intensive for non-specialists
Occupational costs become more fixed and predictable, especially if the occupier is keen to manage and monitor space and energy usage	There may be locations with uncertain land lease terms that hinder longer term strategy of buying vs leasing
Leasehold – Advantages	**Leasehold – Disadvantages**
Short lease means a business can remain agile and flexible	Changes to international lease accounting may mean that a lease does not "free capital" on the balance sheet as much as it used to historically
Lease costs are somewhat protected with rental laws (vs pre-rental escalation law)	Fit-out costs are typically borne by the tenant so represent a frequently reoccurring significant cost (especially if lease lengths are short) and the business moves frequently to different premises
Less risk of being "locked" to a building that may incur unknown costs and obsolescence	Lease costs can still be unpredictable despite being pegged to rental laws (or fixed increases)
Opportunity to test a locality/business in Dubai without a significant capital investment in a new/unfamiliar market.	Inability to tailor the property to the specific needs of the business. For instance, a propensity for the leasing of single (strata) floors makes it hard to find extra space within the same floor/building when expanding the workforce/business

Source: Author's own.

Box 7.3 Buy vs lease: Financial analysis

The firm is looking to evaluate a case on whether the organization should buy or lease a new suitably sized property in Dubai (11,400 sq ft, based on a 100 sq ft per Dubai worker). The same property is available for lease or purchase. Further assumptions are provided below:

Leasing option	Buying option
New Dubai property: 11,400 sq ft Initial rent: AED1,360,000 pa	New Dubai property: 11,400 sq ft Purchase price AED19 million (NIY of 7%, when incorporating transfer fees (see below))
Rent will increase to AED1,500,000 after Year 5, as per a stabilized rent in the RERA index	Commercial mortgages in Dubai are not fixed rate. The analysis will assume finance is available at 7% and the loan will be fully repaid during the ten-year hold period (the maximum loan duration for a commercial mortgage is 15 years) LTV80% (maximum permitted LTV) Opportunity cost of equity contribution at 10%
Agency letting fees 5%. The 10-year lease agreed will likely to be set out as split leases, e.g., an initial 5-year lease + pre-signed renewal at Year 5 (for an additional 5 years). This avoids the requirement to pay 4% fee on the full cost of a 10-year lease and is common practice in Dubai.	Transfers fees + Agency fees: 4% Disposal cost: 2% Depreciation can be reflected in the terminal/exit yield at the end of Year 10 (+0.5% to the current NIY, 7.5%)

Disregarded assumptions on both leasing and buying:

- Service charge/maintenance will be charged to tenant. However, it will also be the added if the firm was to own and occupy the space so is an equal amount either way – ignore for simplicity.
- VAT charged on rent and purchase price of any commercial property in Dubai, can be recovered by a VAT registered business (with TRN)

This financial analysis is best presented using a discounted cash flow (DCF). The key analysis would include further assumptions on the following aspects of the DCF (not supplied in the table above):

- Holding period: as mentioned in earlier chapters, I typically apply a ten-year hold period when evaluating via a DCF. Therefore, it would be a more comparable analysis if the property holding period for both options was ten years. Therefore, the buying option includes an assumption on fully repaying the commercial loan after 10 years, with an assumed sale price at the end of Year 10 (see further buying assumptions for more detail).

- Discount rate: a Weighted Average Cost of Capital (WACC) approach would seem most appropriate for corporate decision-making. Therefore, if the cost of debt is 7% @80% LTV and working capital can return 10%, the opportunity cost of equity is 10% @20%. When summing these weighted parts, we get a suitable 7.6% discount rate = (0.07*80) + (0.1*20), rounded to 8% (for simplicity).

Using the table and other assumptions above, the DCF for the leasing option could be set out as below:

Lease assumptions

- Area 11,400   Assume a 10-year lease
- Year 1–5 Rent   1,360,000
- Year -10 Rent   1,500,000
- Agency letting fees   5% = AED68,000Assume fee on the first year rent only

Lease option	Year 1	Year 2	Year 3	Year 4	Year 5
Agency fees	−68,000				
Rent	−1,360,000	−1,360,000	−1,360,000	−1,360,000	−1,360,000
Net cash flow	−1,428,000	−1,360,000	−1,360,000	−1,360,000	−1,360,000
PV @8%	1.000	0.9259	0.8573	0.7938	0.7350
Discounted outgoings	−1,428,000	−1,259,259	−1,165,981	−1,079,612	−999,641
	Year 6	Year 7	Year 8	Year 9	Year 10
Rent	−1,500,000	−1,500,000	−1,500,000	−1,500,000	−1,500,000
PV @8%	0.6806	0.6302	0.5835	0.5403	0.5002
Discounted outgoings	−1,020,875	−945,254	−875,236	−810,403	−750,373

(assumed rent is paid annually in advance)

The NPV for the ten-year leasing period would be **-AED10,334,634**. This represents the leasing cost to the business over the ten-year period (a present value).

The DCF for the buying option could be:

Buy assumptions

- Area 11,400
- Purchase price   19,000,000
- Down payment   20% (AED 3,800,000)
- Opportunity cost of equity at 10% (AED380,000)
- Loan amount   15,200,000
- Interest   7% (annual repayment 2,164,138)
- Transfer/Agency fees   4%+ trustee processing fees (AED4,200) = AED764,200
- Disposal cost   2% (based on assumed future sale value = AED440,000)
- New rent (>Y10) 1,650,000 2% pa rental growth (assumed)
- Exit yield   7.5%   Assumed this can be applied to the future rent at Year 10 of AED1,650,000 to give a future sale price of AED22 million (AED1.65 million/0.075 = AED22 million)

Buy option	Year 1	Year 2	Year 3	Year 4	Year 5
Purchase fees	−764,200				
Opportunity cost of equity	−380,000	−380,000	−380,000	−380,000	−380,000
Mortgage	−2,164,138	−2,164,138	−2,164,138	−2,164,138	−2,164,138
Net cash flow	−3,308,338	−2,544,138	−2,544.1381	−2,544,138	−2,544,138
PV @8%	1.000	0.9259	0.8573	0.7938	0.7350
Discounted outgoings	−3,308,338	−2,355,617	−2,181,090	−2,019,537	−1,869,941

	Year 6	Year 7	Year 8	Year 9	Year 10
Opportunity cost of equity	−380,000	−380,000	−380,000	−380,000	−380,000
Mortgage	−2,164,138	−2,164,138	−2,164,138	−2,164,138	−2,164,138
Future sale price					22,000,000
Disposal cost					−440,000
Net cash flow	−2,544,138	−2,544,138	−2,544,138	−2,544,138	−19,015,862
PV @8%	0.6806	0.6302	0.5835	0.5403	0.5002
Discounted outgoings	−1,731,540	−1,603,316	−1,484,5050	−1,374,598	−9,511,734

The NPV for the buy option (with a ten-year hold period) would be – **AED8,416,748**. This represents the cost to the business over the ten-year hold period for buying the Dubai premise (a present value).

From this analysis, it appears there is a stronger financial case to support buying rather than leasing based off the comparison of both NPVs. However, these types of analysis are highly sensitive and also ignore the significant capital that gets locked into the commercial property when buying. As noted, commercial mortgages are not fixed rate so there could be some uncertainty on financial holding costs during the 10 years. Nor has the analysis looked at any assumed capital expenditure that might be required with owning, though it could be argue this is reflected in the exit yield of +0.5% to represent the property's age at the end of Year 10. That said, it is likely as a business occupies the property for greater than 10 years, the financial case for owning the property is stronger. With the buying option, it might be prudent to run a simple sensitivity on the DCF to test the assumptions made in the initial appraisal. An example of how this might be carried out is shown below:

Buy option –Scenarios	Optimistic	Expected	Pessimistic
New rent after Year 10	2,007,338	1,650,000	1,500,000
Rental growth pa	6%	2%	0%
Finance rate	5%	7%	9%
NPV	−4,663,235	−8,416,748	−10,877,859
Probability	0.3	0.6	0.1
Weighted NPV	−1,398,971	−5,050,048	−1,087,786
"Blended NPV"		AED7,536,805	

From this analysis, the expected NPV (i.e., the sum of the weighted NPVs) is calculated at –7,536,805. If we were to follow the assumptions of the scenario analysis run above, then we might expect buying to be a more favoured option if we felt the optimistic conditions were likely and apply the inverse in a more pessimistic view on market conditions. In the example above, I have applied a hypothetical probability to each scenario and then calculated the NPV based on a weighted NPV. However, it would be prudent to recognize, in practice, we do not benefit from a "blended" return based on assumed probabilities. By applying an analytical framework like this, businesses and corporate occupiers are better informed in their decision-making, with these common forms of financial analysis likely to breed open discussions about the true cost of buying or leasing options.

Conclusions

This chapter has presented a range of information relating to commercial occupancy. It provided an overview of the space requirements of businesses setting up in Dubai, highlighting the key differences in leasing fundamentals and costs that might be found in other international markets.

Corporate real estate strategies are data-driven processes. Global property professionals and surveyors need to ensure greater consistency in the information they collate and provide clients. Any advice provided needs to be robust and based on credible information. Yet there are challenges in terms of the true comparison between transactions available in the market (headline rent vs net effective rent, for instance). The data available for businesses to support a decision to occupy commercial real estate were set in the specific parameters of the case study provided. I presented and discussed ways in which a company could measure the performance of their occupancy and improve both cost and space efficiency. These measures exist to principally show whether a business is being efficient and effective in managing a significant cost to operational real estate. The benchmarking shown in the case study analysis helps answer specific questions about the way in which the real estate costs are being managed or could be improved. There follows an inevitable purpose to this type of analysis as it provides a way to improve profitability, if an organization can reduce or better manage its real estate costs. However, one must recognize that there would be other ways in which a business could evaluate these decisions.

The growth in businesses owning their operational real estate prompts the favoured question of whether it is better to buy or lease. In short, the basic freehold vs leasehold decision has several strategic, financial and operational factors which will vary from company to company. These decisions can be influenced by the market timing (expansion or recession); capital invested in the business directly vs property; changes to accounting practice; as well as the sector the organization is part of that controls the importance of owning or leasing property. I have shown a second type of analysis a firm might consider running when deciding whether to buy or lease, based on a DCF approach. In business, this is a well-known method from which informed decisions are made. I have also highlighted the fact that a decision will not be the same for every organization and there is a wide range of non-financial factors that would also need to be considered.

The next chapter progresses on to discuss information and data availability in Dubai.

References

CBRE (2019) Global Prime Office Occupancy Costs 2019. https://www.cbre.bh/en/research-and-reports/Global-Prime-Office-Occupancy-Costs-2019.

Colliers (2019) Global Occupier Guide: Dubai. https://globaloccupier.colliers.com/emea/dubai/ (Accessed December 2019).

Deloitte (2021) Doing Business Guides. https://www2.deloitte.com/gz/en/pages/tax/topics/doing-business-guides.html.

Golan, M. (1999) 'The own v lease decision – Myth and reality', *Journal of Corporate Real Estate*, 1 (3), pp. 241–235.

Haynes, B. P. and Nunnington, N. (2017) Corporate real estate and asset management: strategy and implementation, Oxford: EG Books.

IPMSC (2013) 'Jones Lang LaSalle endorse the formation of IPMS coalition' https://ipmsc.org/2013/11/04/jones-lang-lasalle-endorse-the-formation-of-ipms-coalition/

RERA Rental Index. https://dubailand.gov.ae/en/eservices/rental-index/rental-index/#/.

Weatherhead, M. (1997) *Real Estate in Corporate Strategy*, Basingstoke: Macmillan.

8   Property data

Real estate market information including data about transactions in the real estate market are publicly available in Dubai with differing degrees of accuracy and usability. Investors rely on real estate data to guide investment decisions. Without reliable transactions or historical returns indices, prospective international investors are impeded in their ability to make rational investment decisions, thus become reluctant to venture into opaque markets, and if so, apply higher risk premiums. This makes them much less likely to invest. Access to information on the investment characteristics of commercial real estate markets varies greatly from country to country. In Dubai, the availability of supporting data and transactional evidence has improved greatly over the last 20 years. This chapter is seeking to start by explaining these evolutions and provide commentary on how these might impact property professionals in Dubai.

Historic data availability and performance measurement indicators

The property market in Dubai, as well as many other global locations, has historically suffered from severe information constraints. Historically, property information in Dubai has been provided at the national and city-wide levels in the form of indices, while local data was more limited. Transactions were often analyzed by applying an industry-standard discount to the asking price, which undoubtedly contains an erroneous approach. The published national and regional indices tend to be only produced as a single composite measure for all-property and for the four broad property types (residential, offices, industrial units and retail). Data at the local level and in a more disaggregated form was available at a cost.

In the early development of property portals, REIDIN was by far the largest provider of property data with its direct link to the Dubai Land Department (DLD) title registration information. They provide a range of data information services as well as a composite sales and rental index. In the early years, data providers used monthly samples of offered and list prices (asking) which undermined the true definition of market value, suggesting

DOI: 10.1201/9781003186908-11

also that asking prices had an influence on perceptions of value. That said, the early indices established themselves as the leading benchmark against which most valuers measured their professional judgements. Although transactional data were available through DLD, it came under scrutiny with the lack of title registration that took place pre-2010, with the record of actual transactional evidence being rather scarce. Title registrations related to deals undertaken some years previous to the issuance of a title deed had an impact on the reliability of the information offered to valuers and wider investors before 2010.

In addition, many professional firms have constructed their own indices monitoring within key asset sectors. However, with each of these, there were limitations as they individually only represented a small proportion of the market or in fact, their underlying assumptions excluded certain market participants. For instance, the Colliers House Price Index (HPI) was based on mortgage data through collaboration with key banking/financial providers which, on the one hand was advantageous, allowing 'market value' transactional evidence to be provided. Nonetheless, a lesser proportion of transactions are mortgaged in this market. Therefore, the index excluded cash sales that more commonly (and perhaps disproportionally) had taken place in the market. However, such indices were merited for bringing new forms of property data to the market, addressing the significant issue of data scarcity. With the lack of price information, through infrequent trading or title registration, very little information was made available on the finer details of the transaction. This made the construction of transaction-based indices problematic as well as limited the ability of a valuer to obtain true comparable information. In a broader context, the infancy of a historical timeframe of property information also hindered valuers, simply speaking Dubai did not have enough price and transactional history behind it (the time series of reliably recorded property transactions was too short). In the case of pre-2012, the majority of transactions were private and seldom did these get published. Historically, most valuers relied on available indices (while managing their limitations) as well as sporadic information gained from external agents and investment teams. Since then, changes in the provision of data have been improved significantly. This evolution of property data will be discussed throughout the remainder of this chapter. But first let me explain the importance of data transparency.

Why is transparency of data important?

A globally recognized benchmark for real estate transparency is that produced by Jones Lang LaSalle (JLL), an index that now spans over the last 20 years. Dubai is the most transparent of the 15 markets covered across the Middle East and North Africa (MENA) and now ranks in the transparent category. There are several reasons why Dubai scores better. The relatively well-developed legal and regulatory framework is one factor, with

Real Estate Regulatory Agency (RERA) being widely acknowledged as the best-in-class real estate regulator in the region.

Importantly, progress has been made in the MENA region, although many of the region's property markets are still lagging that of the rest of the world. As with the Asian financial crisis in 1997–1998, the outcomes of the global financial crisis have seen a development towards greater transparency over the last 15 years in Dubai with, an "enhanced" level of investor services provided. This began with, for instance, property companies issuing improved annual report transparency, interactive websites and regular analyst meetings to enhance investment appeal. The importance of this increased information transparency by listed property companies is widely acknowledged (Brounen et al., 2001). Markets with more transparent data are able to attract international capital as it allows for more performance measurement; benchmarking and risk management activities. The importance of performance data is, therefore, fundamental in driving towards improving both market efficiency and subsequent inward investment flows as well, on a practical note, an essential tool for local valuers.

Since 2012 RERA has put in place measures which should further improve transparency. These include collaboration with international bodies, broker certification, complaints process, certified valuation workshops, market data, mediation committees and project review tools. Despite these initiatives, Dubai did suffer a slight decline in transparency between 2012-2014, due to the market still being in the early adoption phase of many laws and regulations, a substantial proportion of which, at the time were not fully defined or deeply understood. This was further evidenced by the lack of participation from institutional investors (as discussed in earlier chapters). These organizations typically dominate real estate purchases in other global markets but are relatively underrepresented in the Middle East region. A fundamental reason for this is the high-risk premium that these investors apply to assets to compensate for the relatively poor levels of market transparency. The lack of accurate market data on demand, supply and other market fundamentals has also been a major factor in creating the oversupply that many sectors of the market experienced. By improving property data Dubai can expect more institutional capital. This was evidenced by the recent investment of $5 billion by the Canadian Fund (CDPQ) into Dubai's core logistic infrastructure DP World Jebel Ali Port and Freezone, demonstrating international capital willingness to enter Dubai. Pricing in the real estate market is very much dependent on the quality of information within the market. Information efficiency provides investors with confidence that they are buying at a "fair price". Property market agents and advisers typically "trade" information and provide buyers and sellers with specialist and expert knowledge on a particular section of the market. Property data in Dubai have been historically very challenging to come by, making valuation and investment decisions somewhat challenging. Key observations from my PhD research that surveyed local valuers in Dubai found that:

- Market data was presented in their raw form and provided a lack of depth and detail to assist valuers to draw suitable conclusions.
- Valuers use transactions from third-party data providers and the data still needed to be screened and put through some form of data validation.
- The lack of institutional-grade commercial assets impedes on the relativity of transactions. Subsequently, price signals and market information does not readily enter the market. Although commercial transactions do take place, the deal is often hidden, making comparative property information, like yield and specific lease terms hard to analyze.
- Any institutional or government action which serves to make knowledge better or more readily available, such as CoStar in the UK, is likely to be beneficial when examining market efficiency.
- The timing lag on data is shorter than most markets but does not provide key information regarding the property description (floor height; building name; unit type). Instead, these are inferred from the raw transaction data, based off a valuer's specific knowledge of the subject building.
- The investment method of valuation is very much the spear-headed approach to property appraisal in Dubai and thus drives the future values of the market. Accordingly, real estate prices and rent growth assumptions are central to the pricing of real estate in Dubai, at least in the short term, given the prevalence of the capitalization method applied in the market.

However, developments in this area have been fast-paced since the publication of my PhD thesis, most notably including the launch of a residential and commercial property index in 2019 and 2021, respectively. In 2018, DLD announced that it will begin to compile more building-specific data on a wide range of metrics for both freehold and non-freehold areas, a further positive step in the development of more transparent market data. This information is now being linked to other areas of real estate practice, such as leasing and property valuations. Furthermore, whilst property transactions have always been made publicly available, in 2022 these were released as an open data source, free to download in a format that is much more serviceable for property professionals (see later in this chapter). As the property market evolves, matures and even enters into a new reality of digital innovation and AI in real estate, we can expect to see better data and information continuing to enter the market. This chapter brings together recent examples from local data providers to showcase what information we now have available to make informed property decisions in Dubai. Furthermore, it will highlight what is being done to this data to improve how clients can use it.

Market maturity and transparency

Up to this point, previous chapters have discussed a range of property-related practices that impact the operational efficiency of Dubai's real estate market. Maturity is a study of relativity and competitiveness (the ease of "doing business") and its those practices whether it be professional practices; standardization of property rights (and protection of those rights); transactional demand; data transparency and presence of foreign investment that control market maturity (D'Arcy and Keogh, 1999). A mature market is more open and as such attracts higher levels of foreign and institutional investments. A city that lacks a commitment to the maturity framework (see Table 8.1) presents a risk to investors and often a risk they are not willing to take. Maturity is related to the evolution of market systems and this is not a time-dependent process it can be linked to proactive governance and a willingness to improve processes and efficiencies. Maturity is a relative process and its definition can be considered dynamic. In order for a market to run efficiently there must be a free flow of information. However, even a mature market is likely to remain relatively informal and decentralized. In these circumstances, the flow of information and the availability of specialist advice become important. The transmission of market information becomes the preserve of professional networks, perhaps with one or more professional bodies regulating the quality of service. One key element of market maturity is therefore the adequacy of an information base. This might range from general qualitative commentary to more rigorous econometric quantitative analysis. Furthermore, research may be expected to increase in status and consist of both practitioners and academics in more mature markets.

Keogh and D'Arcy (1994) identify six factors that are indicators of real estate market maturity (see Table 8.1 overleaf). In terms of some more explanatory details these include:

- A mature market should provide investors and users with a diverse selection of property products so individuals can tailor their property rights to their specific needs. The development sector will also become specialized in order to provide this broad range of products. At its simplest level, the market should cater for the separation of user and investor rights through the creation of licences or lease contracts. The market should also provide for the subdivision of the legal interest in a property into smaller lots. For example, in the form of sub-letting for the user market, or the creation of unitized and securitized investment vehicles.
- A mature market will have the mechanisms to enable demand and supply to respond rapidly and be flexible in both the short and long run, so that market participants can react effectively to new information and opportunities.

- Mature markets will have extensive information flows so that participants are informed of changing market conditions. The market will also be well researched.
- The most open markets in spatial, function and sectoral terms are generally more mature. The greater the level of openness, the wider the information available to outsiders. This facilitates information flows throughout the market and allows market participants to operate across boundaries.
- A mature real estate market will contain a sophisticated property profession with a quality of service regulated by codes of practice and/or laws.
- The more mature the market then the more market practices and property rights are standardized.

Alongside the academic work of Keogh and D'Arcy, JLL publishes a Global Transparency Real Estate Index (GRETI) every two years. This index has been argued as being a more sensitive approach to assessing property market maturity (Akinbogun et al. 2014). Whilst this index incorporates some elements of Keogh and D'Arcy's maturity characteristics, its objective as an investor's toolkit, means that it extends to some of the wider investment attributes of countries. That said, the JLL transparency index is a relative index based on the assessment against a range of investment criteria, rather than reference to any stage in market development. Table 8.1 provides my analysis on the synergy between Keogh and D'Arcy's maturity framework and the JLL transparency index.

Defining market maturity in Dubai

My PhD research examined property market maturity in Dubai and provided commentary on the issues faced by property professionals. Key recommendations were based on better availability of data and accuracy of information and more open and standardized property rights and practices. In order to put the maturity framework in context, Table 8.2 draws some comparisons between London (mature) and Dubai (emerging). The side-by-side comparison highlights some key areas of differentiation and as such establishes some expectation that Dubai practitioners are faced with a greater challenge when it comes to valuation practices or strategic advisory roles. I have added an additional column on whether there have been noticeable improvements in these areas over the last five years.

The real estate market in Dubai in 2008, like many other global markets, was hit with a vote of no confidence and many of the developers who entered on the back of high price growth and market opportunity in the preceding years, left projects incomplete or derelict. The lack of consumer protection at the time hurt many investors who seemingly lost out on the gamble of high returns from off-plan sales. Since then, regulators

Table 8.1 Characteristics of a mature property market

Principal aspects of Keogh and D'Arcy maturity framework	Characteristics	Relevance to the JLL transparency index
A market's ability to accommodate a full range of use and investment objectives	Existence of a well-developed investment market environment: • full range of investment objectives • diverse demand of occupiers for space • developed investment culture • no burdens of ownership	Market fundamentals
Flexibility in a market's adjustment in short and long term	Effective property trade and market actors' ability to react to new information and opportunities	Regulatory and legal Performance measurement
Existence of sophisticated property profession with associated institutions and networks	A market's regulation and professional market players' practice	Regulatory and legal Transaction process
Extensive information flows	Transparency level of the market	Performance measurement
Market openness in spatial, functional, and sectoral terms	Allowance of market players to operate with no boundaries	Market fundamentals
Standardization of property rights and market practices	Role for local property market culture	Regulatory and legal Governance of listed vehicles

Source: Waters (2019).

have looked to proactively support investor protection and improve transparency, establishing escrow accounts to protect investor deposits, issuing off-plan title deeds ("Oqood") as well as more open channels of dispute resolution. These developments have undoubtedly risen investor confidence and demand for local investment. Property registration is also more developed as previously many investors were wary of their security of tenure and system of registered freehold titles. In addition, land registration data availability has improved with more public dissemination of transactions available. In April 2022, DLD released via its open data portal all the property transactions in Dubai (both current and historic) which undoubtedly will bring both greater transparency and more investor confidence to Dubai. While a good move towards greater transparency, it is likely to bring a disproportionate benefit to the residential market versus the commercial market, based on the fact that commercial property requires more specialized appraisal and valuation techniques

Table 8.2 Comparison of mature (London) and emergent (Dubai) market

Characteristics of the maturity framework	UK (London)	UAE (Dubai)	Dubai (2022 update)
A market's ability to accommodate a full range of use and investment objectives	Full range	Limited	Moderate, with new forms of indirect property investment options
Flexibility in a market's adjustment in short and long term	Relatively flexible	Relatively flexible in terms of new supply	Similar, though developers since 2014 have been more cautious on project launches
Existence of sophisticated property profession with associated institutions and networks	Specialized	Professional licensing and presence of local RICS offices	Wide range of certified courses in valuation and agency
Extensive information flows	Good	Limited/opaque	Good
Market openness in spatial, functional and sectoral terms	Open	Limited	Limited in commercial sector
Standardization of property rights and market practices	Good	Moderate	Moderate/Good

Source: Waters (2019), updated by the author in 2022.

(see Chapter 9). One key advantage for the transactional data in Dubai is the speed at which it is made public. A user can search transactions from the day before, unlike the UK public registry which would normally take 3-6 months before the transaction is made public. This arguably makes Dubai a more dynamic data set and less prone to the historical data inefficiencies referenced globally.

The limited range of property products is the main restriction on Dubai's evolution. The level of foreign involvement in the market represents its level of openness is also important. The freehold laws of 2002 relating to foreign ownership in designated areas allowed the creation of strata title as an alternative form of tenure for commercial space. It represented a significant step towards the development of a modern market because it increased the range of property products available to the market and created more favourable conditions for smaller occupiers. It enabled the market to be more flexible to accommodate demand and made the market more effective in meeting diverse requirements of market participants.

The evolution of market transparency in Dubai's real estate market

This sub-section will highlight the key developments that have taken place in Dubai from 2004 to 2022 in relation to market transparency. The JLL GRETI is based on a combination of quantitative market data and information gathered through a survey of the global business network of JLL and LaSalle Investment Management. For each market, the survey data comprise of a collection of both qualitative (75 out of 139 scoring factors largely based on a Likert scale scoring) and quantitative (64 out of 139 scoring factors largely based on internal and third-party data) measures. The JLL index is made up of a number of real estate transparency sub-indices, namely; investment performance; public company performance; market fundamentals; regulatory and legal. Within their classification there is a range of outputs to include; opaque; low-transparent; semi-transparent; transparent and highly transparent. Markets are then assigned to one of five transparency tiers, based on a composite score (Tier 1: Highly Transparent (1.00–1.69); Tier 2: Transparent (1.70–2.45); Tier 3: Semi-Transparent (2.46–3.46); Tier 4: Low Transparency (3.47–3.97) and Tier 5: Opaque (3.98–5.00)).

Table 8.3 summarizes the JLL scoring components and makes note of how these core areas of the index can impact valuation practices.

Dubai's commercial property market history is relatively short compared to developed markets, however, the JLL index does show some significant progress related to its maturity level. The earliest report in 1999 excluded Dubai and therefore the historic analysis can only go as far back as 2004, when the UAE was included. In 2010, there was a further separation of the UAE into UAE (Dubai) and UAE (Abu Dhabi) as they have differing real estate governance. Table 8.4 (overleaf) summarizes the developments made in global transparency rankings for Dubai since 2004.

Despite the importance of a country's overall rating within the index, the market fundamentals sub-index appears most relevant to real estate data. This sub-index examines the availability of time series information on major data including; supply, demand, vacancy rate, rent and yield for offices and other investment properties. On further examination, many countries rank highly across the overall assessment, but few have a data coverage with a time series. In the context of Dubai, this can only become a reality on the establishment of quality data sources which currently has a relatively short temporal scale, only covering a period since 2008. However, now established, data are improving, and this area of transparency should be generally predictable. Earlier reports noted other global markets improving transparency largely through the improvements of information provision, stimulated by a widespread global introduction of REITs and more active cross-border capital flows. This trend is still maturing in the context of Dubai.

Another key area of significance is that of performance measurement. However, in Dubai, these have been sporadic or in-house and as above,

Table 8.3 JLL global transparency index, components

Core area	Sub-themes	Valuation specific areas
Performance Measurement (25%)	Direct Property Indices Listed Real Estate Securities Private Real Estate Fund Indices Valuations	Independence and quality of third-party appraisals; Use of market-based appraisal approaches; Competition in the market for valuation services; Frequency of third-party appraisals
Market Fundamentals (20%)	Existence and Length of Time Series on a range of property data and coverage (location and individual buildings)	Rents; Take-up; Vacancy; Yields; Capital values; Investment volumes; Property transactions
Governance of Listed Vehicles (10%)	Financial disclosure and corporate governance	Accounting standards; financial reporting; public share of real estate market
Regulatory and Legal (30%)	Regulation Land and Property Registration Eminent Domain/ Compulsory Purchase Real Estate Debt Information	Contract enforceability; Land registry; Availability of land registry records; Availability of time series data
Transaction Process (15%)	Sales transactions Occupier services	Quality and availability of pre-sales information; fairness in bidding process; professional and ethical standards

Source: JLL website, 2022.

their reliability towards transparency will only really come into fruition after they have been available for a reasonable period of time. In Dubai, one of the key barriers has been the administration of property transactions and timely title registration. However, in recent years this has been significantly improved. According to the 2016 report, the Dubai government continued to develop innovation around improving transparency which has seen an increase in the levels of foreign direct investment. The sharing of information between public and private stakeholders (open data legislation); standardization of real estate processes and contracts and the prompt resolution of real estate disputes have aided the improvement in its global transparency rating. The JLL index points to the fact that regulation and improvements in performance measurement

Table 8.4 Summary of JLL transparency ratings in Dubai (2004–2022)

Year	Rating	Transparency	Global ranking	Synopsis of major changes
2004	4.31	Low Transparency – Tier 4	45/50 (90.0%)[a]	1st data recorded.
2006	3.77	Low Transparency – Tier 4	44/56 (78.6%)[a]	No significant change.
2008	2.78	Semi-Transparent – Tier 3	32/82 (39.0%)[a]	A booming market led to greater transparency (through higher transaction volumes and calls for improved legislation from investors). In 2008, Dubai recorded the largest improvement in transparency, moving up one full tier.
2010	2.93	Semi-Transparent – Tier 3	37/81 (45.7%)[a]	The 2010 report noted a slight deterioration in transparency levels, being one of several countries to record a drop in transparency. Dubai has, however, also taken the lead in introducing important regulatory reforms that have the potential to improve market transparency over the next few years. Positive notes include collaboration with international bodies, broker certification, complaints process, valuations workshops, market data, and dispute resolution committees.
2012	3.05	Semi-Transparent – Tier 3	47/97 (48.5%)[a]	In 2012 the JLL report began to introduce sustainability measures to the index and the lack of developments at the time may have been a reason for the drop in the rankings, alongside the continued addition of new countries.
2014	3.11	Semi-Transparent – Tier 3	49/102 (48.0%)[a]	Dubai that has featured among the top improvers in previous surveys appeared to have lost some impetus between 2012 and 2014.
2016	2.9	Semi-Transparent – Tier 3	48/109 (44.0%)[a]	Transparency is generally improving in MENA, led by Dubai, which has made further progress but remains in the "Semi-Transparent" group.

2018	2.79	Semi-Transparent – Tier 3	40/100 (40.0%)	Dubai is one of the global top improvers, with a host of government initiatives, enhanced regulatory procedures and an increasingly dynamic proptech sector including a building classification project covering all buildings in the Emirate, improved regulatory procedures, new and enhanced online apps for managing contracts and broker information, and unified lease forms. It is also benefitting from additional market data as REIT structures become more embedded and third-party data providers expand their output (e.g., REIDIN, Property Monitor).
2020	2.75	Semi-Transparent – Tier 3	36/99 (36.3%)	Made steady improvements, but need to address issues around corporate governance and regulatory enforcement if they are to progress into the "Transparent" tier. Dubai is leading trials in the use of blockchain. Launch of new official transaction-based residential price index in 2019 (Mo'asher). As sustainability forms part of the metrics, Dubai has Green Building Regulations.
2022	2.56	Transparent – Tier 2	31/94 (33.0%)	Dubai becomes top improver. Introduction of new sales and rental indices for commercial and residential market. Digital platform REST covering a range of online real estate services.

Source: Summarized from JLL GRETI 2004–2022.

a Relativity of global ranking over time, the lower the % the higher the relative global ranking (author's own).

were most significant during the 2008 reforms whereby improvements included:

- Freehold law allowing foreign investors to purchase land/property in pre-defined "freehold" areas (security of title).
- Market regulation through the creation of RERA, particularly in terms of dispute resolution between landlords/tenants and developers/investors.
- Formal sales registration through the DLD.

According to the JLL transparency index, Dubai has been classified as the top performer in the Middle East region. Whilst Dubai in 2016 was ranked 48 out of 109 countries, it scored relatively well in relation to the listed vehicles (26th out of 109, 1.9) but poorly in relation to investment performance (46 of out 109, 3.4). This suggested there were still areas of development needed to improve the ability of professionals to fully evaluate investment performance; value and risk. Since 2018, Dubai real estate has made some notable improvements in relation to real estate data and performance measurement. These include the introduction of two new property indices:

- Mo'asher Residential Index, introduced in 2019 (DLD/Property Finder)
- Commercial Property Index, introduced in 2022 (DLD/JLL)

In 2022, Dubai was the top improver in terms of the JLL transparency rating. Major changes in Dubai were the introduction of new sales and rental indices for commercial and residential markets as well as the digital platform REST regarding property transacting and other property-related digital services (rental increases, service charge management, automated valuations and transactions databases). DLD has dramatically increased the availability of property market information with a full suite of transactional information (see below). Trends in property prices have begun to be charted using time series data which, in turn, can help a valuer; investor and end-user gain confidence and consistency in valuations and appraisals. One key benefit now is the fact that the time series of property price information in Dubai now extends beyond ten years, meaning any analysis becomes more meaningful and significant sample points improve transparency. A new building well-being/sustainability metric is also being introduced. Despite major improvements to move into "Transparent" tier for the first time since the GRETI was launched, investment performance and sustainability (building energy reporting; energy performance and climatic risk reporting were measured) are areas of future improvement. Dubai is likely to see a move from provision of data to deeper analysis as we look ahead.

What real estate data are currently available in Dubai?

Historically, most property datasets in Dubai were unstructured but as the city has evolved so has its reporting of data. Whilst all property sales transactions are recorded via their registration with DLD, a range of competitive data sets has been produced over the last 15 years to assist the local property profession further. Within the investment, valuation space Dubai has three main "external" data providers. These are:

- REIDIN
- Data Finder
- Property Monitor

These three subscription-based providers have historically been very popular as they cleansed the raw DLD data and provide more structured information that has been seen to improve transparency.

Agency portals, where data are being added to support a residential buying decision include:

- Property Finder
- Bayut
- Houza
- DXBInteract

These types of external data have become more popular, with general market indices typically provided for free, and micro-level transactions generally via a paid subscription. In addition, large global consultants do provide market analyses on a range of their specialisms. Since the inception of a number of local property indices, practitioners have been able to track the performance of property. Until very recently, indices were restricted to the housing market and used to track general movements in the Dubai residential sector. In 2020, it was announced by the DLD in collaboration with JLL, that a commercial index is now launched. Figures 8.1 and 8.2 show the current indices available via DLD open data indices, both of which track the performance of those respective market sectors.

Box 8.1 explains the value of a property market index and the challenges presented through its construction.

Property transactions are freely available to search from the DLD webpage, however, there is a requirement to screen what is freely available via this portal. Over time this data source has become much clearer and there is now some depth in what information can be viewed from DLD directly. Real estate transactions can be viewed via: https://dubailand.gov.ae/en/eservices/-real-estate-transaction/#/. Users are provided with daily, weekly, monthly data summaries that show the number of transactions; the total value of these transactions and a brief breakdown of what types of transactions were completed during the time period (separated as unit; building and land). The user can filter and search this information for both residential and commercial transactions. Users can view property transactions either as a daily or monthly summary. The benefit of this government portal is that the

Figure 8.1 Residential sales price and rental price index.
Source: created from Dubai Land Department, 2022: https://dubailand.gov.ae/en/open-data/
indexes-home/#/.

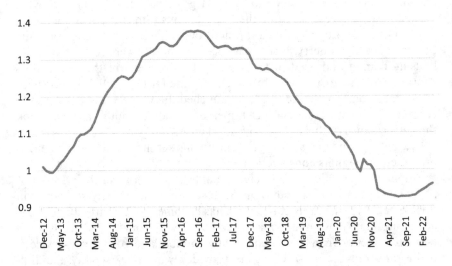

Figure 8.2 Commercial sales price index.
Source: created from Dubai Land Department, 2022: https://dubailand.gov.ae/en/open-data/
indexes-home/#/.

Box 8.1 Reading between the lines: How useful are property indices?

The global property market has historically suffered from information constraints. Where the information on prices in other financial markets, such as equities, is openly displayed and monitored on a real-time basis, the flow of information on property transactions often relies on agent contacts and the trade press. Dubai's real estate market is set to benefit from even more property data and information flows. In 2020 DLD and JLL signed an MOU to create a "Commercial Property Price Index" (CPPI) – but what exactly is a CPPI and how useful are they to the profession and wider property market? The surveys in my PhD, which studied the valuation variance on commercial property in Dubai, pointed to the need for more information efficiency; including more yield evidence; transaction comparables and standardized property data record-keeping. My PhD found that greater market transparency through the pooling (and publication) of property data would ensure more consistent valuation advice. But their impact goes much wider than the valuation profession itself. Property indices guide the investment decisions of firms and large institutional investors as well as offer a central transmission channel of new property information. In the property market, indices are commonly used as estimates of market activity and to benchmark the performance of property against the performance of other investment classes. They are used as measures of volatility and market return, and to estimate the risk and diversification characteristics of property assets. Used in this manner, indices are fundamental aids that enable investors and fund managers to make informed decisions when they are constructing multi-asset portfolios.

In order to construct a CPPI, differences within the transactional data need adjusting and there are a number of academically standard ways of doing this (hedonic regression; repeat-sale; appraisal based) but each method carries with it a series of its own limitations. Users of global property indices need to ensure they understand how the index has been constructed and its merits to gain maximum utility from the new information it provides.

Property indices can be further classified by what they measure. The most common types of indexes include:

- Total return indices (rental growth + price growth).
- Rental or rental growth indices.
- Capital or capital growth indices.
- Yields.

The exact measurement methods used in the construction of indices vary from index to index. If you are using indices make sure that you know how the sample and measurements are defined and calculated. This is particularly important if you are comparing different indices or comparing portfolio performance measurements with figures from an index. Remember, when you are using indices (or any performance measures) always make sure that you are comparing like with like. In Dubai, we do not yet have a comprehensive total return index, yet rental and capital growth indices do exist.

Against this backdrop, I would be advocating the need for data consistency across the profession to improve and be more consistent. DLD is instrumental here and can continue to play its role in ensuring the profession is recording a standardized range of information when it comes to property transactions. The huge advantage is that DLD is able to control both the quality of data and its consistency, benefitting greatly from what the newly announced CPPI can bring to the Dubai real estate market. Personally, I am very excited about what new information (and opportunities) this brings to Dubai's real estate sectors, not only assisting real estate valuers but also the wider market.

information is current as of the same day; therefore, there are no time lags in information recorded as we observe in other markets, the UK for instance suffers a 3–6 month time lag. In addition, commercial property transactions are also recorded in Dubai via the DLD portal.

If you wanted to focus on a particular building for the purpose of an appraisal or valuation, the webpage allows users to search these transactions and see specific sale price information. When searching for villa/townhouse transactions the user must refine the search and change "Property Type" from unit to building. This allows users to search for specific villa and townhouse sales within Dubai. Let us assume that we are looking to find transactions in March 2022 for Burj Views in Downtown Dubai, we can use the portal to search for transactions in that period (1st March–31st March 2022). A summary of this output is shown in Table 8.5.

The range of values in Burj Views for March 2022 for property transactions are from AED1,060–1,245/sq ft. This would give the valuer a starting point from which to make a suitable comparative analysis.

Moreover, users can also search for historical transactions from the DLD Open Data feature on the same page. Under "Open Data" navigate to "Real Estate Data" and a user can search a historical custom range of property transactions. These searches can then be downloaded into Microsoft Excel where the user can search and filter out the specific property transactions more easily. Within this database, the DLD has included a good range of additional information that would be useful for a property valuer and

Table 8.5 Sample transactions from DLD Open Data platform (example: Burj Views)

Project	Property	Sale sequence	Building	Total area	AED/sq ft
Burj Views	Flat	Second	Podium	595.03	1064.63
Burj Views	Flat	Second	Tower B	727.00	1237.96
Burj Views	Flat	Second	Podium	654.02	1223.20
Burj Views	Flat	Second	Tower B	915.05	1202.12
Burj Views	Flat	Second	Podium	734.00	1062.67
Burj Views	Flat	Second	Tower C	1487.05	1244.08

Source: DLD Open Data Transactions, Accessed March 2022.

wider professional, including: transaction date; number of bedrooms; floor level; parking; nearest metro station; nearest mall. The user would need to be familiar with the district from which the property is located which is straightforward and can be searched via public web-based directories. Let us continue with our evaluation of Burj Views. Burj Views is an Emaar development located in Downtown Dubai. It is classified as being in the "Burj Khalifa" area. Historical transactions in Burj Views can then be searched and viewed from 2010 onwards. For the purpose of a valuation, the user would search the transactions from a specified period, in this case, the last 3 months (January 2022–March 2022). From this, depending on the subject property type (studio, 1 B/R, 2 B/R), the valuer could select the most suitable property size. Furthermore, there is an indication of the floor level (and potential views) that would impact value. Being in Excel there is a range of further analysis a user could apply to draw down on specific data points such as the average sale price for a 2B/R and/or an average price per sq m (sq ft) for 2B/R in the building. This portal, therefore, offers users a very comprehensive data set from which to draw out a comparative property analysis. This is a marked improvement to the level of information that used to be disclosed by the DLD and demonstrates Dubai's willingness to improve real estate data and transparency.

The case example below shows how a valuer could use the DLD Open Data Platform to construct a comparable evidence matrix.

Case example: Using DLD transactions to value a residential apartment in Dubai

The valuer will receive the title deed from the owner/client and this document provides information on the property. Key information that would be used from the title deed on the subject property is shown below (for illustrative purposes).

• Community: Marsa Dubai ("Dubai Marina")
• Building Name: Bay Central

- Area Sq Meters: 107.89
- Floor No. unspecified (in practice this information is on the Title Deed)

Using the transactions from DLD in Excel, the valuer can firstly select the community's name ("Marsa Dubai", Dubai Marina). Next, the valuer is likely to review the property transactions within the same building (in this example Bay Central) and seek to find identical units in the building (107.89 sq m). This provides a list of historical transactions. Of course, the valuer is only interested in recent transactions, so would search 2021–2022. This provides two recent transactions of identical unit sizes within the same building as the subject property. The first transaction is for AED1.225 million in September 2021 and the second property transaction sold for AED1.83 million in January 2022 (see Table 8.6).

The valuer might look to calculate an AED/sq m from other recent 2B/R transactions to support these two sources of comparable information. From this, the average AED/sq m sales rate is AED14,311 psm could be applied to the subject property (AED1.55 million). It may also be prudent to widen the search for similar-sized 2B/R apartments in Marsa Dubai in close proximity to Bay Central West, but these two comparables will form a strong part of the valuer's advice and feature top of the list when putting together a hierarchy of evidence. When searching wider, the DLD transactions identify seven buildings (Jumeirah Gate, 5242, Marina Wharf, Marina Quays, Sunrise Bay and Beach Vista). After a cross-check/property inspection, Marina Quays is the second closest comparable building to Bay Central both in finish and location. Jumeirah Gate (superior finishes) and Beach Vista/Sunrise Bay/5242 (new build, beachfront property) and Marina Wharf (poorer specification/finish) can be removed from the comparable matrix. From transactions in similar-sized 2B/R units, the valuer would see transactions of AED1.8 million and AED1.83 million in Marina Quays (in March 2022). This substantiates the evidence provided above and confirms the single transaction recorded in Bay Central in 2022.

Table 8.6 Comparable transactions for subject property

Transaction date	Area	Property	Amount	Size (sq m)	Rooms	Project
01/09/2021	Dubai Marina	Flat	1,225,000	107.89	2B/R	Bay Central West & Central Towers
14/01/2022	Dubai Marina	Flat	1,830,000	107.89	2B/R	Bay Central West & Central Towers

Source: DLD Open Data Transactions, Accessed March 2022.

The data from DLD suggest that the subject property could be valued within a range of AED1.8–1.85 million, with greater weight being given to the direct comparable transactions within Bay Central versus the additional transactions listed above. An indicative value based on this analysis for the subject property could be AED1.83 million. This case study highlights the improvements that have been made to provide more transparency to the market. The introduction of the DLD Open Data portal will aid both valuers, investors and homebuyers alike. The final section draws on the importance of understanding how property data are typically collated and what those numbers mean in practice.

How are averages reported in real estate?

Most analyses in real estate are trying to provide an average rate per sq ft or an average residential value. On face value, an average provides a suitable reference point for property professionals. However, there are three different statistical definitions of an average and the way they present property information is slightly different. Below is a short definition of each of the three averages: mean; median and mode.

A mean average is perhaps the simplest form of average as we are summing the data points and dividing them by the number in our sample. This method includes everything and does not provide exclusion of the "outliers" in a list of transactions. The implication of using a mean average would be that it can make the information reported (i.e., growth or decline) more volatile. If we have a large price transaction in one quarter versus the next it will disproportionately increase the average sale price in that location. In the case of Dubai, luxury homes that sell on The Palm will increase the average significantly if we were to be calculating the mean average house price on The Palm. However, price portals do make a separation between apartments and villas in most cases and this removes the impact of this to some extent when using a mean average on property market analysis.

Another way to remove the impact of a small number of high-value transactions would be to use the median average, which is the middle value of the data set. Though this average is less useful with small datasets, which may be the case in some neighbourhood analyses. A median is best suited to construct a city-wide analysis. The most well-known property data portals in Dubai, such as Data Finder and Property Monitor would typically use a median average.

The final type of average is a modal average (or "mode"). This average calculates the most frequent value. In real estate transactions, this can be a challenge as often a recorded transaction is varied by small differences in agreed prices. This makes the ability to calculate a modal average somewhat challenging. Prices would need to be rounded to the nearest hundred thousand dirhams for this to be a simpler task, even then it becomes difficult to evaluate the most frequent price without some significant change to the data. This type of statistical output would work well in development feasibilities as it provides an indication of the neighbourhood/district's real estate purchasing power. They are less meaningful when trying to provide a property valuation or track price movements over time.

To put this in some context let me calculate these averages against a number of recorded property transactions in the villa community of The Springs (over a 12-month period, April 2021–April 2022).

Number of transactions	*49*
Mean	2,467,143
Median	2,400,000
Mode	1,700,000

In summary, the use of simple averages like these is suitable for comparative analysis and property valuations when the analysis is very localized. The challenge comes when there is a lack of data available to draw meaningful conclusions from. Property indices, however, would find little use in any of these averages as they do expose the information presented to some notable flaws, the greatest being that an average price rise may not be attributed to a real increase in the market, it may simply be an outcome of buyers purchasing higher value homes in that period or quarter. Therefore, a property index would typically look at property characteristics via a hedonic model addressing one key limitation of simple averages not being able to consistently track a property market, as transactions on similar properties infrequently occur.

Conclusions

The chapter has pinpointed the evolution of property market information and data in Dubai. Pricing and decision-making in the real estate market are very much dependent on the quality of information. In 2022, Dubai was classified as a transparent market for the first time since the inception of the JLL global transparency rating 20 years ago.

The data available in the market today are much better at equipping investors and buyers a clearer idea on price performance. It also gives banks an opportunity to supply more innovative finance products as Dubai's transparency and timely data provision reduce the uncertainty around pricing debt-backed assets, like real estate. Despite signs of increasing interest rates in 2022, bank margins on lending should reduce because of greater investor confidence and market maturity bred from significant improvements in data transparency.

The case example of the DLD Open Data platform showcases how purchasers can now evaluate the prices prior to purchasing property in Dubai. This portal as well as a range of other private sector-led offerings referenced in this chapter make decision-making more efficient and in an economic sense closes the gap of price equilibriums between buyers and sellers. This, in turn, reduces the inherent risk premiums that were historically priced into Dubai's (opaque) market. I would expect to see this breed more institutional investment into Dubai as well as promote more residential end-users and investors to make rational purchase decisions.

/Yet when compared to our global benchmarks such as the MSCI (former IPD) datasets, Dubai provides very little on a portfolio level (measuring investment performance), limited to tracking either a residential sale price or a commercial lease price movement. Commercial yield data are still opaque and at best global consultancies provide a yardstick on what investors are seeking at a single point in time (as shown in Chapter 5). In terms of the residential sale price indices, these have largely drawn on average transaction prices. With the regulation of all leases (both commercial and residential) through the Ejari system being a mandatory requirement, it will no doubt breed more innovation around this space and provide users of property information with the much-needed separation of detail in these data sets. Again, the DLD Open Data real estate portal contains rental transaction data that can be viewed and like the sales data, freely exported to Microsoft Excel, for filtering and detailed analysis.

The next chapter focuses on how real estate is valued in Dubai, and the environment for valuers given the discussions within this chapter on property data.

References

Akinbogun, S., Jones, C., and Dunse, N. (2014) 'The property market maturity framework and its application to a developing country: The case of Nigeria', *Journal of Real Estate Literature*, 22 (1), pp. 217–232.

Brounen, D., Scheweitzer, M., and Cools, T. (2001) 'Information transparency pays: Evidence from European property shares', *Journal of Real Estate Finance*, 18 (2), pp. 39–49.

D'Arcy, E., and Keogh, G. (1999) 'Property market efficiency: An institutional economics perspective', *Urban Studies*, 36 (13), pp. 2401–2414.

Dubai Land Department (2022) Real Estate Data. https://dubailand.gov.ae/en/open-data/real-estate-data/#/.

Dubai Land Department (2022) Real Estate Transactions. https://dubailand.gov.ae/en/eservices/real-estate-transaction/#/.

JLL (2022) Global Real Estate Transparency Index 2004–2022. http://www.jll.com/GRETI.

Keogh, G. and D'Arcy, E. (1994), Market maturity and property market behaviour: an European comparison of mature and emerging market, *Journal of Property Research*, 11 (3), pp. 215–235.

Waters, M. (2019) A critical examination of property valuation variance in Dubai, PhD thesis.

9 Property valuation, methods and techniques

The internationalization of property standards and legislation has been a focus for Dubai over the last 20 years, partly driven by a similar globalization of financial reporting, international accounting standards, and increased volumes of international trade in real estate assets. Property valuation is an area of professional practice that entails the estimation of value (or most likely selling price) of property assets. Alongside market transparency, valuation accuracy and standards impact investor confidence, particularly in relation to institutional investment. An inability to provide credible professional advice on the assumption that valuations are a good proxy for prices would also impact market efficiency. Dubai has significantly improved valuation standards and regulation, much of which conforms to International Valuation Standards (IVS). This chapter aims to provide a background of information related to valuation governance in Dubai. An overview of valuation methods and techniques along with a range of worked examples is then discussed. The chapter concludes by looking to the future of the valuation profession and highlights what innovations are imminent. The first section of this chapter begins by examining the main valuation regulations present in Dubai.

Valuation regulation in Dubai

Valuers and RICS firms in Dubai raised concern during the global financial crisis over the absence of fixed rules for the evaluation of property assets. They argued it was one of the main factors that resulted in a deterioration of investor confidence. Nowadays, property valuations are a systematic process for a suitably qualified professional to assess the likely exchange price for an asset (or liability) at a specific point in time. Dubai government have two main regulatory bodies in the property market. These are:

- *Dubai Land Department (DLD)* was established to act as Dubai's official registry, valuer, auctioneer, regulator, information provider and property "gatekeeper". Both buyers and sellers use this institution to record officially all transactions and transfers of ownership. They also serve as a public information source for property market information and sales/rental transactions.

DOI: 10.1201/9781003186908-12

- *Real Estate Regulatory Agency (RERA)*, with main objectives to monitor and regulate the Dubai real estate market, including the regulation, managing and licensing of various real estate activities. Currently, there are 50 registered firms that are licensed to undertake property valuations in Dubai. RERA provides a list of certified valuation firms in Dubai at: https://dubailand.gov.ae/en/eservices/approved-valuation-companies/valuation-company/#/.

Since 2015, under Executive Council Resolution No. 37 of 2015 related to the organisation of the valuation profession, practicing valuers in Dubai must be licensed and obtain registration on the real estate valuation registry at the Real Estate Regulatory Agency (RERA). Practitioners who undertake property valuations have been required since 2015 to possess certain qualifications and be registered with RERA. Requirements for valuer registration in Dubai include (amongst others):

- Valuers who are UAE nationals should have no less than two years' valuation experience. For non-UAE nationals, the minimum requirement is five years' experience.
- Valuation experience should be documented with three sample valuation reports per year.
- Valuation experience can be in any country in the world. However, UAE experience is preferred and recommended (for a minimum of six months). Valuers having less than six months' UAE experience are liable for further checks.
- Valuers should attend and pass the Certified Valuer Training Course in Dubai. Box 9.1 summarizes how to register as a valuer in Dubai.

Box 9.1 Registering as a real estate valuer in Dubai

In order to register as a valuer, the individual must provide the following to RERA:

- Certificate of experience in the field of real estate valuation – two years for citizens; five years for expats
- Certificate in real estate valuation (from IEREI or equivalent)
- A certificate of good conduct and behaviour from Dubai Police addressed to RERA
- Profile picture; copies of passport and ID card. The residency visa of an expat real estate valuator must be on the same licence

---

Once licensed, the valuer will abide by specific rules. The latest rules are set out in the *"Charter for practising the real estate valuation profession"* (1st issue, June 2020). Some notable conditions placed upon registered valuers within this document include (RERA, 2020):

- Objectivity and ethics of the real estate valuation profession: The valuer must issue unbiased reports regarding the reliability of the data and assumptions and to ensure that the valuation process is credible, following a code of ethics for the profession enforced by RERA.
- The valuer should mention their name and registration number in all communication, certificates and reports issued when preparing a work report for each real estate valuation task that is carried out. A summary of this report should also be sent to RERA.
- Preparing a paper or electronic file for each real estate valuation task that is carried out, including a copy of the written reports, communication and notes related to the task.
- Maintaining records, reports and files related to properties that have been valued for a period of five years starting from the date of the end of the task, or from the date of submitting the final report. RERA may request to check these.
- Taking into account the form, basic requirements, and data of the property and the rights related to it, which must be included in the report,
- Apply real estate valuation criteria and methods set forth in the Emirates Book Valuation Standards (EBVS)
- Not to use any person not registered in the registry to carry out the real estate valuation process
- The valuer should practise the profession at only one office in the Emirate

The licensed valuer must adhere to these standards, otherwise as with other global professional bodies, administrative violations; penalties and fines can be imposed (see the above-mentioned *"Charter for practising the real estate valuation profession"* (1st issue, June 2020) to review these in specific detail).

These regulations have largely been introduced to ensure public confidence in the ability of valuers in Dubai to meet a recognized global standard. Additionally, a number of additional standards have been established to support the development towards internationalized valuations. These notably include:

- The EBVS, issued by Taqyeem, provides a framework for valuation standards and methodology in Dubai and follows the standards of the International Valuation Standards (IVS) published by the IVS Council.
- Taqyeem has developed a Code of Ethics for valuers practising in the Emirate of Dubai

• International Property Measurement Standards (IPMS) mandated by the Dubai government, which aims to unify the way property space is measured internationally by reference to a set of consistent property measurement standards. The IPMS had been implemented in Dubai since 2015.

The following sections outline the key valuation standards operating in Dubai in more detail

Valuation standards operating in Dubai

The valuation standards in Dubai are based on IVS, termed EBVS and communicated within the Emirates Valuation Book. A positive element is that this book complies with the main points of all major IVS. The main structure of the EBVS is shown in Box 9.2.

Box 9.2 Internationalization is at the heart of Dubai's property valuations

The structure of the EBVS
PART ONE: The EBVS framework
Definitions; Objectives; General Bases;
Measuring spaces according to EBVS
and International Property Measurement Standards (IPMS)
PART TWO: General standards included in EBVS
EBVS 101: Scope of work (contract terms)
EBVS 102: Verification and compliance
EBVS 103: Reports preparation
EBVS 104: Bases of value determination
EBVS 105: Valuation methods and approaches
PART THREE: Other standards included in EBVS
EBVS400: Property interests
EBVS410: Real Estate Development
                (Source: DLD (2020) "Emirates Book of Valuation Standards")

The EBVS is based on three main international standards:

• International Ethics Standards (IES)
• International Property Measurement Standards (IPMS)
• International Valuation Standards (IVS)

Key standards within the EBVS relate to professional ethics; market value and reporting which are all briefly summarized below. As with other global valuation professions, valuers in Dubai need to manage a range of rules and ethics to ensure the valuations are impartial. A summary of these is related to managing ethical behaviours, such as:

• Standards
• Integrity and honesty
• Conflict of Interest avoidance
• Confidentiality
• Neutrality, transparency and accountability
• Competence and external assistance
• Professional indemnity (PI) cover

Key areas of notable reference include: avoiding conflicts of interest; only undertaking valuation work where they possess suitable skill and experience and producing valuation reports that are clear; transparent and not misleading. The valuer is prohibited from working for two or more parties in the same transaction without obtaining prior written consent (indicated in the report). In case of a conflict of interest, valuers can either: take necessary measures to preserve information and prevent "information/data leakage"; obtain written consent from all parties to move forward or refuse to carry out the valuation instruction. The valuers must refuse any work if they lack experience or skill to accomplish the task efficiently. If the valuer requests assistance from an external party, they must ensure the adequacy of their competence. These follow similar guidance as other international valuation regulations.

Valuation standards are set up to follow IVS and one of the key reference points when estimating value, centres on defining the basis and purpose of a property valuation. EBVS defines the bases of value for Dubai valuers. These are summarized in Table 9.1. The most common basis of value is market value and, in accordance with EBVS, unless noted or instructed otherwise, valuers are recommended to use market value as a basis of value (DLD, 2020).

The EBVS also provides several mandatory requirements for valuers undertaking property inspections in Dubai. These include:

• Obtain owner's consent to visit the site (and tenants if applicable)
• Inspect the property on the ground, indicate condition and prepare a detailed report that includes building systems; building completion certificates; title deeds and leases
• The date of submitting the valuation request and visiting the site
• The type of materials that make up the property and percentage of each material
• The age of the building
• The general appearance of the building in terms of its condition compared to the surrounding real estate
• Determine geographical description of property in terms of accessibility and other descriptions

Table 9.1 Six bases of value in Dubai (as per EBVS)

Basis of value	EBVS/IVS definition	EBVS guidance
Market Value	"The estimated amount for which an asset or liability should exchange on the valuation date between a willing buyer and a willing seller in an arm's length transaction, after proper marketing and where the parties had each acted knowledgeably, prudently and without compulsion."	"EBVS recommends use of this definition for all valuations conducted in Dubai unless there is a valid reason to use another basis of value". In such case this shall be in agreement with the client in advance and documented in the report. "When considering any valuation report in Dubai, the market value shall be the assumed basis of value unless otherwise stated in the report"
Market Rent	"The estimated amount for which a share in real property should be leased on the valuation date between a willing lessor and a willing lessee according to appropriate lease stipulations in an arm's-length transaction after proper marketing wherein the parties had acted knowledgeably, prudently and without compulsion."	Whenever Market Rent is provided the "appropriate lease terms" which it reflects should also be stated. For instance EBVS 104.30.2 states "it is necessary to consider the rent value…in the contract, if different from the market rent". EBVS requires the use of Ejari system when searching for any leasing data. Use of any non-Ejari leasing accounts shall not be considered unless otherwise approved and mentioned in the report
Fair Value	"the estimated price for the transfer of an asset or liability between identified knowledgeable and willing parties that reflects the respective interests of those parties"	It is worth noting that the main difference between FV and MV is the lack of marketing for the property. Also, the price agreed might reflect special advantages or disadvantages that would not be present if the property were to be sold in the open market.

Investment Value	"The asset value for an identified or potential owner in terms of individual investment or operational goals"	In some circumstances, the worth of a building to a particular person is the same as the price it would fetch if sold in an open market. However, this is the exception and not the rule. Investment value typically meets the specific criteria of an individual or organization. Thus, do not confuse Worth with Market Value.
Liquidation Value	"the amount that would be realised when an asset or group of assets are sold on a piecemeal basis"…determined under:	a – an orderly transaction with a typical marketing period b – a forced transaction with a shortened marketing period Dubai Valuers must disclose which premise of value above is assumed
Premise of Value or Assumed Value	EBVS 104.80 adopts Premise of Value or Assumed Value to include: Highest and best use Current use/existing use Orderly liquidation Forced sale	

Source: Summarized from EBVS.

- Real estate measurement according to IPMS (with dual reporting, if required)
- Take notes and photographs on interior and exterior (noting particular defects)

In the event, the property is examined by someone other than the valuer this must be clearly stated in the report (with names, dates, qualifications and job titles (DLD, 2022)). Similar to other professional valuation standards EBVS have devised a set of minimum report contents that all valuers should abide by. The minimum contents are:

- EBVS3.1 Identify the Client
- EBVS3.2 Purpose of Valuation
- EBVS3.3 Subject of Valuation
- EBVS3.4 Basis of Value
- EBVS3.5 Date of Valuation
- EBVS3.6 Status of Valuer: External or Internal Valuer
- EBVS3.7 Assumptions, Special Assumptions and Departures from the Standards
- EBVS3.8 Statement of confirmation with Standards (EBVS)
- EBVS3.9 Opinion of Value (amount in words and figures)
- EBVS3.10 Name and Signature of Valuer
- EBVS3.11 Associated Documents

The EBVS states valuers must maintain an electronic record of each real estate appraisal task, including copies of reports, correspondence and written notes related to this task (must be held for a minimum of five years from the valuation date). For this purpose, of complying with the requirements, EBVS states valuers must comply with following in relation to informing their client of the scope of work:

- The nature and scope of work of the value and any restriction imposed on this work
- The nature and source of information the valuer relies on
- Type of report being prepared (and restriction on use and distribution)
- Any time limit in relation to the valuation (validity applies only to the day of the valuation)
- In case of the EBVS does not address an issue, the valuer should refer to the IVS. If a contradiction exists, then the requirements of the EBVS will be followed

Despite the presence of the EBVS, many valuers also adhere to other global valuation standards such as the RICS Red Book for valuations conducted on behalf of clients. In these instances, the traditional large international valuation consultancy firms are favoured given that they have practising RICS members. Some local valuation companies are also seeing benefit in gaining RICS membership and acting for global clients, who take confidence from the RICS global regulation. Since 2008 GFC, local banks have required secured lending valuations to be undertaken by RICS firms, further supporting Dubai's drive to support international standards. The next section moves on to discuss how valuations are undertaken in Dubai.

Undertaking a basic valuation in Dubai

The previous discussions on the availability of property data and valuation in Dubai draw attention to the imperfections that exist when valuing real estate in global markets that are less transparent. Yet, the same valuation principles are applied to Dubai real estate as they are in other markets. This section will explain how properties are valued in Dubai and provide a local context to how these international methods are applied in practice.

How do we value real estate in Dubai?

The EBVS defines five internationally recognized valuation methods. Each method is suited to particular types of assets for a particular purpose (see Table 9.2). Broadly speaking, the comparative method is the most preferred and accurate when equipped with a high number of good-quality transaction comparables. The grid below summarizes the suitability of each method for market rent and market value reporting. The method used will suit different property assets.

Direct comparison is the simplest valuation method, largely used for single-unit residential villas/townhouses/apartments. It is used in the component part of other valuation methods whereby comparative data are required, such as market rents and yields for the investment method. It can also be used to cross-check land valuations (versus the primary method, a residual valuation). The valuer directly compares the subject property with a number of suitable comparables that have recently sold using historical transactions. A valuer should analyze and make adjustments for any material differences between the comparables and the subject property (often using a comparable matrix – see Box 9.3).

Table 9.2 Key valuation methods used in Dubai

	Comparison	Investment	Residual (Land)	Profit (Specialist)	Cost
Capital value	✓	✓	✓		✓
Rental value	✓			✓	

Source: Waters (2019).

Box 9.3 Case study: Valuing a residential villa in Dubai, The Spring 5 (2,777 sq ft)

You are valuing the subject property for a refinancing valuation (market value, secured lending) in March 2022. You have been provided with an average sales rate of AED890/sq ft and a range of recent transactions. Comparables 1–4 are also of a similar size to the subject.

• What factors would you consider based on the comparable information provided?
• How useful is the information provided to make a comparable valuation?
• The client is recommending a value of AED2,950,000. Do you agree?

In this simplified example, the valuer is presented with a variety of information. First, an average sales rate (AED890 psf) might be used to benchmark a value. In this case example, multiplying AED890 psf by the property size of 2,777, provides a starting point of AED2.47 million. Second, the valuer would examine the comparable transactions. Valuers apply a "hierarchy of evidence" to the analysis which in this case would mean a sale transaction carries more weight than an MOU, which carries more weight than an asking price.

• Comp 1 is the best evidence as it is the most recent sale.
• Comp 2 is a useful reference as the MOU in the current month means a deal has been agreed upon between a buyer and seller (though has not been formally transferred at Dubai Land Department (DLD)).
• Comp 3 as an asking price can be somewhat disregarded
• Comp 4 appears a little outdated though may provide a base level (as another sale).

Lastly, the valuer would take the evidence from Comp 1 and Comp 2 (those deemed most suitable) and establish any adjustments that might need to be made between them. For example, the subject property is located close to power lines which typically are less desirable than those set further in the community. Comp 2 is close to a major road, which might be detrimental in terms of noise, or an advantage based on ease of access in and out of the community. Despite Comp 1 being a better specification, it appears the market has moved upward since that transaction was recorded and the evidence might suggest AED2.7–2.8 million is a suitable value for the subject property. The client's recommended value appears too high to support the market evidence provided.

Property	Type	Beds	Layout	Parking	Location	Price	Evidence	Date
Subject	Detached	4		Garage	Nr. power lines			
Comp 1	Detached	4	Study	Better	Better	2.7 million	Sale	Jan-22
Comp 2	Detached	4		Better	Nr major road (E311)	2.85 million	MOU	Mar-22
Comp 3	Detached	4	Large dining	Better	Double garage	Better	2.95 million	Asking	Jan-22
Comp 4	Detached	4	Large dining	Better	Better	2.5 million	Sale	Nov-21

Source: Transaction prices sourced from Data Finder, accessed March 2022.

Income-producing assets, like shopping malls; offices; industrial units, are valued using the investment method. This method is derived by multiplying the current rent by a yield (or cap rate). Box 9.4 explains this approach in more detail.

Box 9.4 How is value created in real estate for an income-producing asset?

- Real estate value is driven by rent (occupiers) and yield (investors)

- Investor demand and opportunity cost (return per unit of risk) impacts yield

If a property returns a cash flow of AED100,000 pa and investor returns on comparable properties are evidenced at 10%, then the value of this cash flow can be assumed to be AED1,000,000. This is based on a yield multiplier of 10 (100/10) × 100,000 = 1,000,000.

If rents remain unchanged but investors perceive Dubai property to be less risky and are willing to accept a 5% yield, the value of the same cash flow increases to AED2,000,000 (100/5 × 100,000).

Let us take the latter example above as a comparable transaction. An investor purchases the property for AED2 million (including purchaser costs) for the rights to receive an income of AED100,000 pa. Therefore, the initial yield on this transaction would be 5%. The initial yield expressed is an output to the rent, yield and value relationship. A valuer would take this comparative transaction and observe that a rational investor for this type of asset is willing to accept 5%. The yield becomes a point of reference to determine the cap rate multiple (100/5 = 20). Therefore, an income-producing property that provides a cash flow of AED120,000 pa could be reasonably valued at AED2.4 million, including costs (120,000 × 20). In practice, the rental income should be a net operating income (gross rent (or that contracted) minus the operating expenditure). A valuer would often need the input of a property management or asset manager to estimate what the expected operating expenditures are. In markets like Dubai, where the operating expenditure is often passed onto the tenant, assumptions from gross to net income can be relatively straightforward (and would follow the parameters laid out in the lease).

The study of the predominant valuation approach of commercial valuers in Dubai during my PhD research found that many valuers adjusted their practices by adopting "layer" methods (over-rented) or "term and reversion" methods (under-rented), which recognizes that the capitalization of rent can be split into two distinct sections. Taking the example of the over-rented property, the market rent (or bottom layer) is what similar properties are currently renting for, capitalized at the market-derived ARY. The top layer (or overage) is subsequently capitalized at a higher ARY to reflect the uncertainty of the property remaining "over-rented" versus the current Estimated Rental Value (ERV). Box 9.5 provides a typical example of how valuers apply a cap rate to property valuations in Dubai.

Box 9.5 Valuation of a residential building (freehold area)

This instruction refers to the valuation of a single-owned residential building within a popular freehold area of Dubai. The building had a mix of studio and one-bedroom apartments.

Valuation purpose and basis: Secured lending/market value
Valuation method used: Investment method. The valuer chose to use the capitalization method.
Comparable sales of residential buildings in Dubai: Yield evidence of 8%–9.75%. On account of the most relevant transactions against the subject property, a yield of 9% was adopted.

Summary of valuation workings

Annual rent (Fully occupied)		*AED5,400,000 "a"*
Less		
Building services (chiller, electricity, management, maintenance, insurance) "b"	−1,025,000	
Vacancy (@2.5%)"c"	−135,000	(1,160,000)
Net Income		4,240,000
YP in perp @9%	X 11.1111 (100/9)	AED47,111,111
Sale price (net of 4% fees)	−1,811,966	(1,811,966)
Market value		AED45,299,145

Market value reported rounded at AED45,000,000 (net)."a" rent assumed to be net to the landlord with service charge paid by the tenant. "b" utilities for common area not included in the lease. "c" using a % of rent is flawed when the occupancy of the building is far below a normal range. In case of low occupation, an absolute cost figure should be used.

The residual method, used for land with development potential is discussed in detail in Chapter 10. Specialized assets, like hotels and cinemas, utilize the profits method, so properties become valued as business entities and the valuer is primarily concerned with establishing a rental split (i.e., if the business performance is X, what remains between operator/tenant remuneration and property rent). This established rent can then be capitalized with an appropriate yield. This method requires very specialized experience as many of the assumptions drawn out in the valuation need expert knowledge, skills and experience. Similarly, the cost approach (or depreciated replacement cost) requires specialists. The method is based on valuing the land and building components separately with the building replacement costs taking into account property age and obsolescence. For the purpose of this book, the latter examples will be excluded. While the cost approach cannot be used for secured lending purposes (cost does not equate to value), the method is often applied by valuers in Dubai as a sense check against a primary valuation method. Whilst the EBVS refer to the same five methods of valuation as the RICS' Red Book (comparable; investment; cost; profit and residual), there is a preference for certain techniques to be applied in Dubai. Most notably in the investment method of valuation, the EBVS encourages valuers to use a discounted cash flow (DCF) approach. This technique maps out the future net cash flow (income – outgoings) in a more meaningful analysis than capitalization methods. The remaining parts of this section will discuss the use and application of DCF valuation techniques in more detail.

Dubai as a proponent of the DCF method

The EBVS advocates the DCF method as a more appropriate valuation method. The origins of the ARY in valuations was derived during a period when it was valid to assume rents were fixed (a direct result of long commercial leases, and inflation-linked rental growth). But how applicable are these assumptions in Dubai where typically the average commercial lease is less than five years (DLA Piper, 2020). Not only that but the ARY approach was deemed suitable in global markets where valuers have an abundance of comparable rent and yield evidence. As already discussed, a major challenge for Dubai valuers is the sourcing of suitable comparable evidence. So how applicable are the conventional ARY valuation methods and assumptions in Dubai?

The benefits of applying DCF techniques to modern-day property valuations are related to its flexibility (to reflect the more varied income from shorter leases) and the valuer can explicitly take account of their impacts on the risk and income growth (Baum and Crosby, 2008), akin to the US market, where the origins of using the DCF in property valuations and appraisals were first held. The main arguments in support of the contemporary DCF techniques are:

*   Traditional techniques break down in the absence of good comparables so that they often include subjective manipulation of information by the valuer.

- Traditional techniques have in the past produced price inefficiencies, for example, in the short leasehold market.
- DCF-based techniques take a more rational approach to the valuation of the income flow.
- More flexible technique so can deal with short leaseholds and unusual costs and receipts.

The DCF itself is not without criticism. The main complaints refer to the difficulties associated with accurately forecasting future cash flows and the subjective nature of selecting an appropriate equated yield. Advocates of the traditional methods argue that these methods are more objective because the estimation of market values relies purely on comparable transactions. The purpose of a valuation is to predict price. If the market is using irrational methods, so should the valuer. However, even the most ardent supporter of the traditional techniques accepts the need to use DCF-derived techniques where there is a lack of supporting comparable evidence – as such this appears suitable for Dubai.

The fundamentals of a DCF valuation would be the valuer estimating cash flows over an assumed holding period (typically ten years), plus an exit value (often referred to as a terminal value) that is calculated by applying an exit yield to a future rental value (at end of Year 10). The future cash flow is then discounted back to the date of valuation (present-day) at a discount rate that reflects market risk sentiment (evidenced by a series of similar property transactions and their net initial yield). The net initial yield represents the rent divided by the purchase price of the property (inclusive of transaction fees). Box 9.6 explains how a discount rate in Dubai might be calculated. The resulting Net Present Value (NPV), therefore, represents the current value (to either the "market" if the assumptions are market-based as they would be in a valuation or an individual/corporate purchaser if the assumptions are bespoke to a particular client as they would be in appraisals).

Box 9.6 What is a discount rate? How do we define it in Dubai?

The simplest working definition of a discount rate is that it is the desired return or target rate of return. This is assumed to be different for different individuals and firms, based on the principles of understanding an opportunity cost relative to risk. A simple example is if I can hold a relatively secure investment (ten-year indexed linked government bond) at 2% per annum, for riskier assets I would need to receive a higher return to reward me for the additional risk I am taking on the alternative asset. In most global markets, real estate returns or discount rates can be defined from building up a suitable risk-adjusted return from the base point of an index-linked ten-year government bond. Historically, in the UAE we did not have such benchmarks (see Chapter 5, Box 5.3). In the absence of a risk-free rate of return benchmark in Dubai, most practitioners use the cap rate (or initial yield) + inflation as a suitable measure of desired returns (or discount rate). Of course, the discount rate is a term that is interchangeably used by valuers between market valuations and appraisals, yet their definition is different depending on whether the valuer is doing a market valuation or a client-specific appraisal. In the appraisal of an investment or development opportunity, the discount rate can be defined using the Weighted Average Cost of Capital (WACC) approach. I provided an example of how this can be done in Chapter 7 (see Box 7.3). Ultimately, it is a simple measure of opportunity cost of capital vs cost of debt. If my capital (equity) I invest in a property can return 8% elsewhere (say stocks) then I would apply 8% to my equity contribution (say 20%). If the current cost of debt is 3% (on the remaining 80%), my WACC would be $(8 \times 0.2 = 1.6) + (3 \times 0.8 = 2.4)$, then totalled together my WACC would be 4%. This may serve as a suitable discount rate for my specific investment profile. Similarly, developers can use this when appraising land development opportunities and each developer has different forms and access to finance.

The discount rate gives the valuer an indication of what level a "rational" investor is currently pricing risk at on similar grade assets, through the principles that an investor is seeking the highest return for the lowest level of risk. As the cash flows are future payments (based on current lease + future assumptions), they need to be discounted using an appropriate discount rate, derived from market data (typically drawn from comparable transactions in the marketplace or built up from a risk-free rate of return), as shown in Box 9.6. Moving on from this, Box 9.7 shows how a valuation DCF might be set up via a simple example. The output of an NPV represents the sum of the discounted future cash flows generated by the income-producing asset, therefore, would be expressed as "market value".

Box 9.7 Case study: Valuation of a new commercial building in JLT (freehold area)

This instruction refers to the valuation of a single-tenanted commercial property within a popular freehold area. A valuer has been instructed to value a commercial tower in Dubai, located in Jumeirah

Lake Towers (JLT) compromising of 16,500 sq ft. The current passing rent is AED3,700,000 per annum for the first three years after which the lease is assumed to track rent increases per the RERA rental index. The index shows that the current rent cannot be increased as it is less than 10% of the average similar rent. However, a 5% increase is likely at Year 7 (assumed either through the RERA rent index or a new letting based on three-year leases). Operating expenses are estimated to be AED15 psf/annum (escalating at 2% per annum). Recent transactions are indicating the investor's target rate of return (discount rate) of 7.5% and assume an exit yield of 7%. The net income at resale of AED53, 964,986 is assumed as the future sale price based on a future rental net income of AED3,777,549 (AED4,079,250 minus AED301,701) divided by 0.07 (assumed exit yield). A simple DCF for this income producing asset could be set up as below:

Year "a"	Income	Expenditure	Net income	PV @ 7.5%	Discounted net income
1	3,700,000	247,500	3,452,500	1.0000	AED3,452,500
2	3,700,000	252,450	3,447,550	0.9302	AED3,207,023
3	3,700,000	257,499	3,442,501	0.8653	AED2,978,908
4	3,700,000	262,649	3,437,351	0.8050	AED2,766,932
5	3,700,000	267,902	3,432,098	0.7488	AED2,569,957
6	3,700,000	273,260	3,426,740	0.6966	AED2,386,925
7	3,885,000	278,725	3,606,275	0.6480	AED2,336,727
8	3,885,000	284,300	3,600,700	0.6028	AED2,170,340
9	3,885,000	289,986	3,595,014	0.5607	AED2,015,732
10	4,079,250	295,785	3,783,465	0.5216	AED1,973,393
Resale	4,079,250	301,701	53,964,986	0.4852	AED26,183,483
NPV				Market value	AED52,041,921

a As it is typical for rent to be paid annually in advance, the DCF assumes annual in-advance payments.

The resultant NPV of AED52,041,921 is a gross amount and valuer would typically state the market value after deduction of 4% purchaser costs. The net figure is AED50,040,309 (52,041,921/1.04) which could be rounded down to a suitable valuation figure of AED50m.

What is provided in Box 9.7 is a simplified example of a DCF valuation for illustration only. In practice, with an investment valuation using a DCF, there are a number of more detailed considerations that need to be mapped out by the valuer in order to calculate a market value. Below is a summary of these discussions.

***What key variables do you need to consider when valuing a
commercial property in Dubai?***

A range of industry-standard terminology required to complete a comprehensive property valuation/appraisal includes the need to nominate discount rates, capitalization rates, void periods, acquisition dates and holding
periods. In addition, a distinction is made between the current cap rate and
that on disposal (or exit yield). Some of these key areas will be outlined in
brief detail below:

a   *Holding period*
     How long will the property be held in the portfolio? Will it be sold?
When? There is often a sale considered after a rent review or lease renewal. Disposal tends to be difficult before review or lease renewal due
to uncertainty in the future rent. Investors will be more interested once
this uncertainty (risk) has been reduced. In the US, where DCF analysis
is nearly universal, ten years is often assumed as a holding period. This
is subject to criticism since there has been no empirical confirmation of
this assumption; however, it does appear rational as beyond ten years
a valuer's ability to foresee rent escalation and yield adjustment would
become more challenging. Although the property may not be sold after
ten years, it is still valid to assume a notional resale date in order to
assess the expected holding period return. Typically, in Dubai, valuers
also assume a ten-year holding period. If the property is freehold, the
valuer would also need to apply a suitable terminal value (net income
divided by exit yield) to consider the perpetual income under this form
of property ownership.
b   *Future rental income (growth and depreciation)*
     When considering rental income beyond the current lease, the val-
uer needs attention to two key areas, namely; rental growth and rental
depreciation. Modern software packages (like Argus Enterprise) allow users a straightforward and transparent means of examining the
projected rental values by defining annual escalation rates that can be
set against market commentary or standard benchmarks, such as the
Retail Price Index (RPI) or Consumer Price Index (CPI). However, in
Dubai, it would be more prudent to escalate the rent to a stabilized
rental value as per the RERA rental index (as this law governs rental
increases in Dubai). It is important to be as explicit as possible with any
growth assumptions made. The other major issue affecting future rental
income is depreciation. Depreciation refers to the decline in value as
buildings grow older. This would be more apparent during long periods
of low inflation. Hence, with consequent lower levels of rental growth,
depreciation will become a more important variable affecting property
investment returns. So how do we take depreciation into account? One
possibility is looking at different levels of rent for buildings of a different

age which are in a similar location. Alternatively, the valuer could assume a capital expenditure (as a cost) to bring the subject property back up to a modern-day equivalent (see outgoings).

c   *Voids*

There is a possibility (probability) that an existing tenant will choose to relocate to new premises at the end of the lease. Or the tenant may choose to exercise a break clause if this is appropriate. In Dubai, where leases tend to be short, it is likely that the possibility of voids will be an important consideration. A rental void can easily be included in the cash flow as a % deduction of the net income, based on an assumed natural vacancy level for the asset. The presence of either single-tenant or multi-tenant properties would also have important considerations. Table 9.3 summarizes these as a side-by-side analysis.

d   *Outgoings*

There is a range of outgoings that also need to be considered including the cost of acquisition/disposal fees; conveyance fees and necessary rental incentives. In addition, there are capital expenditures that are large single payments of expenditure to maintain the property to a modern-day equivalent standard. Outgoings will also depend on the different lease terms. Properties in Dubai have typically given the tenant a greater responsibility for paying for the operating expenses (via an additional service charge above the rent payment) which means fewer deductions are required off the annual rent, similar to a full repair and insuring lease (FRI) we observe in other markets. Valuers often assume an operating expense for the building for the annual management of services and larger maintenance. Although some state this as a % of the annual rent, a more appropriate method would be to use an AED/sq ft

Table 9.3 Single-tenant versus multi-tenanted valuation assumptions in Dubai

Single-tenant investment	*Multi-tenant investment*
More predictable returns with cap rates between 5% and 8%	Higher cap rates often above 10%
No/low vacancy	Less likely to be 100% vacant at a time
Periodic rent increases built into the lease and in accordance with DLD legislation	Tenants as smaller individual entities often have lower strength in the negotiation of lease terms
Greater risk of rental income gaps if in poor location	Riskier and require more active management
Often longer lease terms	Often shorter, more flexible leases reducing income security
More favourable finance rates for secured lending given more institutional investment qualities	Finance might be at higher rate due to higher risk proposition

Source: Author's own.

annual cost, as the building management costs are not correlated to the occupancy levels in a building (should be viewed as fixed annual costs).

e   *Terminal capitalization rate (terminal yield/exit yield)*

In principle, the calculation of the terminal cap rate at resale is straightforward. But how do we forecast the exit yield (cap rate at resale at Year 10)? Can we use current yields? This seems inappropriate since the building will be ten years older and will have suffered depreciation and obsolescence. Can we use current yields for buildings which are ten years older? This seems more defensible. However, it involves an implicit assumption that current macro-economic conditions will remain at resale date.

Evaluating risk in property valuations in Dubai

Valuers and analysts plot key information regarding the existing tenants of the building and understand the proportion of income generated by different tenants in the case of multi-tenanted properties. In addition, lease expiries and option probabilities can be examined to cast light on the longevity of the current income stream as well as forecast and anticipate future revenues (by making reference to the ERV overtime). Essentially, the user can adopt a series of analyses based on tenant profiling and study the implications if break clauses were to be exercised, or renewal incentives need to be part of the leasing negotiations. These will be based largely on how attractive the current passing rent is to any reversionary income. The valuer should when using a DCF be able to advise the client better and examine their assumptions in more detail. Table 9.4 highlights the key risks within a commercial property valuation.

The key issue concerns risk measurement. The most commonly used quantitative measure of risk is volatility – measured as a variance around the mean. Probability is a simple way of measuring uncertainty, and probability

Table 9.4 Risk classification in a property valuation or appraisal

Risk	Examples	Interpretation
Low risk	Rent passing	The current rent is known and a contractual obligation of the tenant. Where there is a good quality tenant (covenant) this can be considered relatively certain
Medium	Cap rates, current LTV, depreciation expenditure	These data are usually evidenced by transactions so can be estimated with a good degree of certainty. Issues may arise when there is a lack of good comparables
High	Growth rates, rental escalation, future income and costs	Forward-looking variables are the most uncertain/risky variables. Forecasts can be made but these are likely to contain some degree of error

Source: Waters (2014).

is used to describe the amount of uncertainty present. As discussed earlier having an appreciation that today's lease structures are shorter and contain break clause options (early in the lease), a valuer does need a mechanism whereby the likelihood of the current tenant renewing or breaking their lease can be made. Modern-day software packages like Argus and Estate Master do allow for a range of scenario and sensitivity testing. The use of scenarios would be more appropriate to an appraisal rather than a valuation (unless it was perhaps done for loan security purposes).

Using scenario testing, it is possible to examine the outcomes of the NPV ("market value") under different economic circumstances. The Expected Net Present Value technique simply develops this analysis further. A probability is assigned to each scenario. In this way, each scenario is "weighted" and the Expected Net Present Value can be calculated and cross-tabulated to a separate user sheet. The basic method is to group the various estimates to suit particular circumstances or scenarios. For example, if a user wanted to test their rental growth assumptions, they are able to use Excel/ software to model the below-mentioned scenarios:

Scenario A   Deteriorating economic conditions
Scenario B   No change in the economic environment
Scenario C   Improving economic conditions

Scenario	*Optimistic*	*Expected*	*Pessimistic*
Rental growth	7.50%	4.5%	2.50%
NPV	AED2,325,962	AED1,269,961	AED651,206
Probability	0.3	0.6	0.1
Weighted NPV	697,789	761,977	65,121
Expected NPV ("market value")		AED1,524,886	

This simple scenario testing gives a valuer the opportunity to factor in changes to the market and wider economy when providing advice. The DCF technique does permit this form of analysis. The advantages of such an approach are that it overcomes the limitations of a sensitivity analysis in that the likelihood that certain combinations of market conditions occur is taken into account; it enables the valuer to examine the impact of different sets of circumstances. It, therefore, encourages the explicit assessment of the probability. However, the limitations would be on the subjective elements of defining a suitable probability. Such techniques are more commonly applied to appraisals and development feasibilities.

On the assumption that these and others are applied correctly, DCF techniques permit the valuer the following results:

1   They force the valuer to make decisions in a logical and consistent fashion with as much quantitative and qualitative precision as possible. By having a standardized framework that is easily modified with the

specifics of the particular income-producing asset, an extensive analysis can be modelled when required.

2   The DCF allows the valuer to be much more specific about the criteria on which decisions are made. In addition, it offers a consistent approach to the analysis and evaluation of acquisitions and disposals at the portfolio level.

3   It provides a standardized framework and transparency.

When applying the DCF to real-life valuations or investment opportunities, a user will be able to see the merits of a more systematic method to the analysis, allowing a consideration to "one-off" considerations, especially those dealt with under the conditions of uncertainty. Any model used to evaluate investment opportunities should be able to include an expression of the uncertainty and risk associated with variable factors of an investment – which most significantly would include three core elements: rent passing (income) over the holding period; discount rates and the yield (or cap rates). We are all familiar with the purpose of modelling – and that is for us to enable a problem to be studied, analysed and adjusted in order to arrive at the most feasible (most often profitable) solution. I believe that a DCF method offers a comprehensive analytical framework and more importantly recognizes the current theoretical trends in property investment techniques globally. Furthermore, specific user requirements can be modelled (via Excel or other spreadsheet interfaces) to reflect the ever-complex institutional structures developing within global commercial real estate.

This chapter will now move on to examine some of the key differences in valuation approaches between Dubai and other international markets.

Key differences on the valuation approaches between international markets and Dubai

The following section will highlight the practical approaches to valuation practice in Dubai, based on a range of traditional income-producing assets. Specialized valuations, such as the profit method used for hotels and other business-revenue properties, while a recognized method in Dubai, has not been referenced in order to keep the discussions relatively simpler and more straightforward. The main aim of this comparison is to inform non-specialists in Dubai real estate valuation an opportunity to appreciate how a practitioner is using a range of valuation methods in their client advice. Each case example and a suitable methodological description are shown below.

Case example 1: Valuation of residential or office building (freehold vs non-freehold area, single-owned, full tower)

Freehold areas: These are locations in Dubai where unrestricted perpetual ownership of property interest is held by an individual or a company.

- Single residential units: In these locations once a project is launched, units are then sold to individual investors on a strata basis. If a valuer were asked to value 1–2 single for internal or mortgage purpose – the most common method of valuation would be a comparable method.
- Bulk residential units: If there are a collective number of residential units being held by an investor, the usual market practice is to value the property on both comparable and investment. These are then considered to be income-producing assets for which the investor is looking to benefit from the rental income. This approach is considered to be preferred by financial institutions and banks as well, because it captures the investment aspect of the property. Further, in the situation of a default, this method provides a view to the bank on how long they are able to hold the property to earn rental income to offset the losses from the default. However, most banks also request to include within the reports, a sense-check based on the comparable method as well. Considering local market practice – although an investment property there are times when an investor might offer to buy it based on a rate per sq ft rather than purely based on a rental potential and yield return. Especially since the lease terms in the region are relatively short when compared to other established markets such as the UK. And the comparable valuation can be the aggregate value of bulk units valued as at the valuation date or a more preferred way of reporting it as a single investment asset and hence take off a portfolio discount to account for the time required to sell all the units.
- Full residential building: If a building is held on a single-title deed the approach is very similar to the above of the bulk units valuation. In this case, a discount is almost always applied to the aggregated market value to arrive at the value of the whole building. Non-freehold areas: Market practice is always to value adopting the investment method, as they are often held for investment purposes. Most of these buildings are owned by private local families and businesses and are held on a single title, rather than strata. The historical areas have always had a good rental demand and benefitted from higher occupancy rates. These are especially preferred by occupiers looking for affordable accommodation as well as by ones who are working in nearby offices and commercial locations. Valuers are always mindful of the land prices in these locations. Since traditional areas are highly varied in terms of asset types, location and quality (with some of the older buildings in Dubai situated in very prime locations), the standard rules for land vs building value go out of kilter in these locations. In some rare cases, valuers have also experienced, redevelopment potential and the land value being higher than the existing use/investment value. In these cases, the valuer would report the higher value (i.e., the land value in this case).

Case example 2: Valuation of industrial property

The majority of industrial assets are held on leasehold titles in the older areas. Few locations such as Al Quoz or Quisais can have a mix of freehold and leasehold. Usually, in these areas, the method adopted is the investment approach. However, valuers will always keep an appreciation of the value of the land in these locations when undertaking valuations. Sometimes a one-year rolling lease is granted in these areas and then while valuing you must adopt certain special assumptions otherwise one-year rolling leasehold interest will practically have no value.

Areas such as JAFZA and DIP are considered very popular industrial areas in Dubai. The warehouses here are held on a leasehold title ranging from 10 years to 90 years. DIP used to previously grant a 90-year leasehold to a lessee; however, now they grant 30 years with or without an option to renew. These should be mentioned in the ground rent agreement. Most of the warehouses in the location are owner-occupied and can significantly differ in terms of quality and specification. Hence, as market practice when these assets transact, they usually do so on a rate of AED per sq ft on the BUA or GIA. Valuers in the region value these assets on a comparable rather than on an investment method. There are very few warehouses that have been subleased because of the sublease costs and restrictions to do so (see Chapter 6). Also, when these transact and the lessee is changed, the leasehold does not get transferred to the new purchaser (on a number of years remaining on the lease basis), instead, a completely new lease is drawn with 15 years or 30 years (as applicable).

Case example 3: Valuation of retail property

Retail and commercial, unless it is a small strata unit on the ground floor of a building is almost always valued on an investment method. The only challenging part about valuing retail and commercial assets (like malls or buildings) is the assumptions made around the continuity of the lease (for example, lease breaks, rent reviews, treatment of service charges).

How do Dubai property valuations compare to other global markets?

There is a vast number of academic studies that have examined property valuation practices in other global markets. Table 9.5 identifies the three main categories of global valuation studies.

My PhD research was the first study of its kind in Dubai. The research examined valuation variance in Dubai and benchmarked the local valuation profession against other global markets. Key findings included:

- Around 70% of local practitioners valued the subject property within a 10% range of each other, in line with comparable international studies.

Table 9.5 Definition of valuation accuracy, variance and bias

Key term	Definition
Valuation accuracy	The difference between a valuation of a property and a target price, such as its subsequent sale price (exactness)
Valuation variance	The difference between two or more valuations to produce the same outcome, a measure of the difference between two or more valuations on the same subject property (consistency)
Valuation bias	The measurement of consistent over- or under-valuation of property

Source: Waters (2019).

- Within the Dubai survey, valuers were also able to increase their consistency amongst the group when given identical market information (84.6% within +/−10%).
- Whilst variance measured well when compared to the range of international comparative studies, the results do indicate that local valuers are asked to make judgements based upon a paucity of information.
- A lack of property transactions had a noticeable impact on the ability of valuers to provide consistent yield evidence (which would remain a threat to the variability of commercial property valuations in the local market). The lack of transactions also meant valuers are facing a challenge to enforce reliability, accuracy and objectivity. The central issue of managing valuation variance, however, was reported as being a well-understood mechanism.

I put forward the analytical framework that valuation variance (and error) is an inherent part of any real estate market and the factors controlling valuation variance can be defined under "non-systematic" factors and "systematic" factors. The levels of valuation variance can be removed only through a range of non-systematic factors. In essence, the presence of such factors develops as the maturity of a country's real estate market also develops. However, there is a natural margin of variance that cannot be reduced due to the inherent characteristics within the valuation profession (for instance, the subjectivity of real estate valuations as well as determined by property asset characteristics; heterogeneity; time lags in market information). This margin may be referred to as systematic or non-specific risk and is created by the general characteristics of property assets. These general factors will have an influence on the level of valuation variance on all property assets (to a varying degree, ranging from small for residential to much larger for more complex valuations, like hotels). It is therefore somewhat suggested that variance is shown to be defined by a range of local variables such as the extent of available information; variability of property cycles and the heterogeneity of the property stock.

Table 9.6 What factors impact property valuation variance?

Factor or attribute that impacts property valuations	Level of risk	How is the Dubai valuation profession addressing these?
Professional ethics	Low	A full and complete range of professional licencing and ethical standards (as per EBVS)
Information efficiency	Moderate	Data transactions are accessible and timely yet more opaque for commercial property transactions
Market transparency	Low/ Moderate	Transparent as rated by JLL Global Real Estate Transparency Index (GRETI) in 2022, yet commercial property data is still opaque versus residential
Valuation regulation	Low	Full adoption of IVS/RICS standards and EBVS
Standardization	Low	Full range of standard market practices

Source: Author's own.

Table 9.6 goes on to highlight the broad categories of market factors and practices that may affect valuation variance. The table also shows the extent to which Dubai's valuation profession is managing these valuation variance risks, to which non-systematic factors are a greater source of valuation variance. From this evaluation, information and market transparency remain the most likely sources of variance and valuation uncertainty.

A key criticism of valuers in Dubai during my study was the lack of market information and data. The World Economic Forum (2015) attributed Dubai's real estate volatility to a lack of transparency stating that both lenders and investors had opaque market data (both statistical and transactional) from which to base decisions.

Recent market data suggest that more stable market conditions now exist. At the macro-economic level, this has been attributed to both better real estate regulation and data provision across the industry (refer back to Chapter 8). New regulation such as the RERA enacted law on permitted rental increases provides valuers with a clearer roadmap for a property's future cash flow.

Box 9.8 examines the perceived risks that might be present when valuing in Dubai and how these can be suitably managed. These observations are based on the research I have undertaken into managing valuation variance in Dubai.

Box 9.8 Managing "valuation risk" in Dubai

Historically, valuers have referred to a lack of transactional data in Dubai to undertake valuations. Therefore, there was a danger that valuers conform more to market expectations than data-driven objectivity. However, with the introduction of the DLD "Open Data" platform in 2022, valuers should now verify information from this central source, reducing the influence of third parties. The issue becomes more pertinent as Dubai valuers are often finding themselves working in a new and unfamiliar location.

Figure 9.1 demonstrates that the behavioural aspects are far more intended than those based on market timing or "random aspects". The right side of the diagram illustrates that the behavioural influence comes from three main components: the individual valuer; the client and the wider peer group. Each of these sectors places an amount of "objectivity tension" on the valuation process, for instance, an individual valuer will be influenced to some degree by the client (forced by the potential retention of business revenue) as well as influenced by the wider pool of valuers who look for reassurance amongst each other on key aspects or information guidance (i.e., how closely am I aligned with my competitors or the leading valuation firms?).

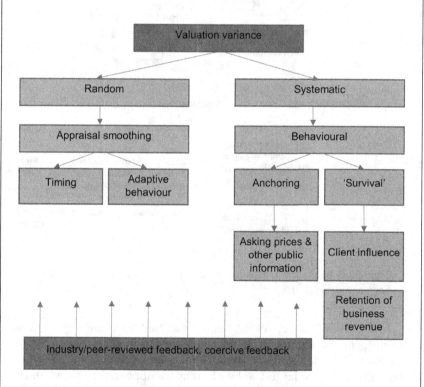

Figure 9.1 Feedback variables creating valuation "risk" (variance) in Dubai.
Source: Waters (2019).

DLD will begin to compile more building-specific data on a wide range of metrics for both freehold and non-freehold areas. This is seen as a positive step. At present, it appears that most valuers rely on available indices (while managing their limitations) as well as sporadic information gained from external agents and investment teams. Looking ahead, more pooling of new information brought to the market via DLD's open-data platform will improve consistency of valuation practices.

A valuer will pass judgement on market value from three data sources; normative data (what should be); positive "hard" data (from transactions) and supporting "soft "data (wider market commentary). As more positive data are available via DLD there is evidence to support greater consistency amongst valuations (see Figure 9.2).

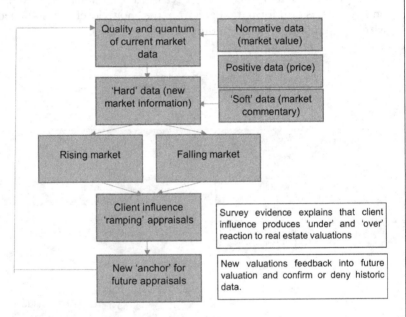

Figure 9.2 Behavioural aspects related to real estate valuation in Dubai.
Source: Waters (2019).

Over 90% of survey respondents noted that they had experienced some form of client pressure. A range of client influences included; valuation negotiations (adjustments within ±10%); opinion shopping and fee retention. The largest body of influence appeared to come under behavioural bias. Valuers disclosed clients promising large value contracts if they are "happy" with the current instruction, or conversely would appoint a competing firm if unsatisfied with the value. As with other market misbehaviours, such as proving collusion between firms in an oligopoly, client influence on property valuations is regarded as taking place, however, it would need proving on a case-by-case basis. Therefore, the valuation profession needs to ensure a consistent approach to manage client expectations and uphold the profession. Survey respondents did suggest some useful remedial action, including pre-payment for valuation work (removes fee retention behaviour of client); a separation between valuer and client; as well as more strict fiscal penalties and sanctions. The unified set of valuation laws and enactments via EBVS are positive steps to improve market practice and consistency amongst Dubai valuers. A further recommendation could be a random allocation of valuation work to firms from clients so that valuers are impartial and not mindful of the sales of valuation services. Random or independent allocation of valuation work would also manage the challenge of fee-cutting in the local market, which is another source of market pressure that exacerbates the bargaining power of the client.

In short, the central theme of these discussions shows that information is key. Valuers must be equally informed and follow equally similar processes and methodologies if intra-valuer variability is going to be contained. This can only happen if information and processes are standardized. Further challenges exist as valuers' interpretations of the same information can be different. A by-product of Dubai's highly diverse expatriate population is that the profession has a myriad of valuation terms, methods and analytics that can also lead to greater variance. My research has suggested that there is a high level of consistency among valuers despite their origin and education/training diversity. Notwithstanding the noteworthy consistency amongst valuers, property valuations in Dubai appear to suffer from two main sources of variance. The survey findings reported these as differences in the quality of current information (transactions) and evaluation of future risk (yield), with the latter still needing improvement. Chapter 5 provided an analysis on how this might be addressed with the introduction of long-term UAE Treasury bonds.

The main recommendations from my 2018 research focused on three key areas of valuation practice, including:

a   Market information
   • Research should be undertaken to establish ways of improving the level of market information shared amongst the valuation community. This would ensure that valuers are less likely to be of different

opinions when it comes to the transactional evidence supplied on valuation work.

- A common international language of valuation terminology needs to exist so that data and information can be collected, stored and shared in a consistent format.
- Research into the interpretation of market value definitions needs further testing.

b   Valuation practices and methodologies

- Research should be undertaken to investigate whether there are any potential mechanisms within current valuation practices to contribute to improving valuation variance further. This would ensure the methods used by local valuers are "fit for purpose" and market risk is more consistently represented.
- Monitoring of valuation methodologies/processes in Dubai should be widely encouraged by regulators and global professional bodies to include those not RICS qualified. This would ensure the valuation industry is operating on a "level playing field".
- More understanding/application of local legislation and governance of real estate. Explicit assumptions on legal interests and lease conditions are a fundamental component of the valuer analysis. More information is needed in the public domain to improve working knowledge of the local laws related to real estate. The traditional perception of real estate as "bricks and mortar" needs to change to reflect real estate as a financial and legal asset.
- Greater consistency in valuation reporting and should evaluate whether risk scoring would allow the profession to be more explicit (and consistent) about property risk in Dubai.
- Develop a universal standard valuation report to improve transparency and improve understanding amongst end-users of the valuation reporting.

c   Policy formulation

- Clearer guidelines are needed on the benefits of IVS in Dubai, especially for clients and wider public stakeholders. This would ensure that the process and purpose of valuation work are better understood.
- Local licensing laws on valuers have caused some confusion and although credited to raise the profile of valuation work, the threat of relevant work-based experience was seen as an important area that needs further clarification. A key danger was an oversimplification of valuation processes as well as the administration of what constitutes relevant work experience. The auditing role of the RICS valuation firms was seen as a positive process, but would only be impacting on a portion of the local valuation industry (with the exclusion of this regulatory service for non-RICS firms).
- There needs to be consistency in local regulation to adopt IVS rather than coming up with local standards. Attempts to localise the rules

of valuation are likely to be a hindrance to the reduction of variance between valuations. This has now been enacted as discussed earlier in this chapter

• The profession should look to avoid the provision of a two-tiered valuation market of international consultants versus local practice.

From detailed discussions with valuers in Dubai, there are some common market observations that are stated when evaluating the challenges of property valuation in Dubai. These would include:

• The use of an appropriate valuation methodology. There are some inconsistencies in the most appropriate use of valuation methods used by valuers. The introduction of EBVS and certifications has improved this historic issue.
• A lack of market rental evidence. This has been solved to some extent with Ejari and DLD open data
• A lack of cap rate/yield evidence, particularly for commercial transactions

A key outcome of my research was the consensus view that property data and information needs improving. Property market data should be recorded in property market templates or involve more explicit capturing of transactions and proxy valuations. The valuation report should source the property market data and rate the quality of the information contained within it. Furthermore, the profession needs to be more collaborative. This could take the form of quarterly submission of information from valuation firms to a third party, independent valuation review panel or regular knowledge transfer of standardized and audited property market data. The local valuer surveys and focus groups undertaken supported the need for clarity, consistency and some new initiatives to keep the momentum towards more international valuation frameworks. There is also an opportunity to create a local valuation profession that is more transparent and collaborative rather than competitive in approach. The societal benefits of information sharing and knowledge transfer are more far-reaching.

Since this research was carried out, Dubai has implemented a range of legislation surrounding the governance of property valuations. The latest edition of EBVS encompasses three pillars of internationalization within the valuation profession. These include: International Ethics Standards (IES); IPMS and IVS. With the adoption of these international standards, Dubai has been able to match similar permissible ranges in property valuations (Waters, 2019).

The future of real estate valuations in Dubai

Since 2008 the emergence of RICS property valuations and other forms of standardization, such as the mandatory use of IPMS in Dubai has contributed somewhat to an improvement in the reliability of market information and property valuations. The findings from key literature and my own research surveys suggest that some outstanding questions remain. These include:

- To what extent are valuers considering risk in the valuation process, and how well understood is it as a concept or paradigm by clients/users?
- At what stage in the valuation process do valuers engage with other key stakeholders? How can this interaction be improved?
- What impact do post-valuation discussions with clients have on variance, accuracy and bias?

Looking forward, further research into property valuations in Dubai or the wider UAE/GCC markets are required. The profession can be reviewed and future questions could broadly include:

Market impacts

For example:

- How is risk defined, assessed and communicated in the valuation process?
- How do valuers report on the quality of their property data and information?

Stakeholder engagement

For example:

- How do valuers engage with other stakeholders (including government, agencies, investment teams and local agents) during the valuation process?
- What are the incentives for greater engagement during the valuation process?
- What are the valuers' cultural responses to valuation in a new global market, like Dubai and wider GCC?

Technology adoption

For example:

- How does the valuation industry use and implement automated technologies?
- How do other stakeholders, such as clients, view automated valuation reporting?

Box 9.9 identifies the role technology can play in property valuations in Dubai.

Box 9.9 Real estate in a digital era: Autonomous valuations

The findings of my own PhD research on property valuation variance have shown that international valuations are exposed to human influence and as such variances. The potential deployment of Artificial Intelligence (AI) within real estate suggests a boom in new models for predictive data analytics. The emotionless appraisal of real estate assets is one that may outweigh the drawbacks of human heuristics and the psychological drawbacks of our inability to forecast the future accurately. The speed of such deliverables will be bound by how the industry standardizes the information we store on buildings and leases which are still hugely differentiated across global practices. This is a market where competition will not exist. Instead, the most accurate automated valuation model (AVM) will rise to the top – after all, why would you choose the second- or third-best AVM. Whilst the adoption of AVMs in Dubai is currently not widespread, they are coming. One such example is a hybrid residential real estate valuation product being developed by Cavendish Maxwell and Property Monitor, that focuses on automated valuation solutions for local banks. A key challenge to date has been the combined impact of regulation and the shorter timeseries of data available. Yet the combined experience of a leading real estate consultancy and a well established Prop Tech firm that can bridge these information gaps via its linkage to multiple data sources and algorithms, could see the implementation of a leading AVM soon. The standardization of industry practices will determine how much human involvement will be part of these AVMs. It will also impact asset classes differently, with adoption more likely in residential markets rather than on more complex commercial asset classes. If AVMs are commonly adopted, a world of more instantaneous bank lending and mortgage offers could be put into existence, speeding up the buying and selling process for property assets. Consumers would also expect mortgage processing fees and valuation fees to reduce as a result of automation. For Dubai as a whole, AVMs will continue to improve efficiency and transparency in the market.

Source: Author updated his original article Waters (2020).

Conclusions

Dubai Government is putting firm legislation in place to ensure a greater emphasis towards attracting institutional investment. A move that should be praised and merited. In an environment of more focused regulation, the local valuation profession has become more rigorous. Looking forward, a range of new government initiatives are being introduced that will further assist valuers. For instance, DLD is likely to release a new version of the rent index after the completion of the building classification survey. Under the classification survey, each building in Dubai's non-freehold and freehold communities will be given a star rating depending on its location, amenities and sustainability factors.

The introduction of more rigorous data streams, more effective modes of communication and improved training opportunities have given a boost to the valuation profession and maintained momentum towards reduced variance (and greater accuracy). The appraisal sector and valuation profession in Dubai is growing in maturity with good progress being made in the areas of standards, ethics and codes of practice.

Dubai's commercial markets are characterized by much shorter leases, frequently with annual changes in income and in some cases nonrecoverable costs; costs which would vary over time. To value these more complex income streams/assets, methods that project forward the varying income and expenditure patterns are needed, namely, a DCF approach. Globally, valuers must start to examine the specific risks set out in the lease rather than opting to "hide" assumptions under the ARY approach. Despite the ARY being suitable in a transparent market, with a high number of comparable evidence, its suitability in Dubai's more opaque market has historically been called into question. While the capitalization method is still applied in Dubai and many other parts of the world, the RICS recently put out international guidance covering the same, establishing a rethink to how investment valuations are to be carried out in the future. Therefore, it seems apparent that valuation practices will start to shift to a more explicit DCF approach and methodology.

Valuation practices in markets, like Dubai, are understandably changing at a fast pace as alignment with international best practices appears the most accepted endpoint. Additionally, investigation into the cost of obtaining property market data is key, particularly within the commercial sector, where transactions and yield evidence are still sparse. It is, therefore, essential that more empirical investigations are undertaken. These will generate additional and complementary data and insights to inform long-term initiatives to improve valuation practices in Dubai and wider Middle Eastern markets.

References

Baum, A., and Crosby, N. (2008) *Property Investment Appraisal* (3rd ed.), Chichester: Wiley-Blackwell.

DLD (2020) Emirates Book Valuation Standards (EBVS). https://dubailand.gov.ae/en/about-dubai-land-department/emirates-book-valuation-standards/#/.

DLA Piper (2020) Commercial Leases in UAE – Dubai. https://www.dlapiperre-alworld.com/law/index.html?t=commercial-leases&s=legal-characteristics-of-a-lease&q=length-of-leases&c=AE-DU.

RERA (2020) 'Charter for practising the real estate valuation profession' (1st issue), Professional Ethics Charter.

Waters, M. (2014) The Importance of Using Investment Appraisal as both a valuation tool as well as investment management, Investment Appraisal White Paper Series, Estate Master.

Waters, M. (2019) A critical examination of property valuation variance in Dubai, PhD thesis.

Waters, M. (2020) Real Estate in a Digital Era, Property Finder. https://www.prop-ertyfinder.ae/blog/real-estate-in-a-digital-era/.

World Economic Forum (2015) 'Emerging Horizons in Real Estate - An Industry Initiative on Asset Price Dynamics', https://www.weforum.org/reports/emerging-horizons-real-estate.

10 Real estate development processes in Dubai

The purpose of this chapter is to provide a general overview of the real estate development process and procedures in Dubai. It should be used in context with other chapters of this book in terms of thinking about the way in which new supply is brought into the market when compared to other global markets. The chapter will also examine how developers make property development decisions and examine the underlying development risk they perceive and manage throughout the process. The chapter starts by outlining the various functions entailed in the property development process and the possible composition of the external stakeholders who often make up the development team in Dubai. Several case studies reflect on the property development process in Dubai.

This chapter introduces the fundamental component faced by all new development – development control and planning policy. Whilst there are some differences in terms of the administration of planning control in most countries, we can observe the essence of "master planning" and "development frameworks" set in place by the relevant local government authority or municipality. This framework is established in order to control development and promote the "best" and most effective "use" of our limited land resources. Given the presence of obsolescence in property markets, a secondary component of development control is the idea of a change in use. Both primary and secondary development is therefore a control function of planning authorities. This chapter is, therefore, designed to give an introduction to these processes in Dubai. The following discussions will describe Dubai-based practice and summarize by detailing similarities and differences amongst other international development frameworks.

Internationally, planning powers represent an invasion of traditional common law property rights in land. As a consequence of this, many jurisdictions offer rights of appeal against the discretionary decisions of government planning agencies – giving relevant development parties an opportunity to request a review of controversial planning decisions. However, with the presence and inclusion of sustainability in many international development frameworks, planning decisions and negotiations are notoriously complex. As such, unwanted time delays and a lack of transparency in the

DOI: 10.1201/9781003186908-13

planning process are common criticisms of development control systems. Prior to local government intervention, landowners were free to use their land in any way they wished, subject only to obligations placed upon them under common law. Therefore, land could be used for the purpose for which it was deemed most economically viable. Nowadays, the situation has very much altered and in most global economies we see a situation whereby this freedom of ownership has been restricted for both public good and to maintain the long-term interests of the wider community.

Despite the clear intentions of the planning process to manage and allocate land use resources effectively, we often see through history and across a variety of international examples, development pressures often exceed the planning regime. Hence, the more outdated a development plan may become, the less relevant it is to make decisions about the use and development of land and the greater pressure to rely on other material considerations than rely on the development plan. In practice, that has led to many local government practitioners making land use decisions on an individual and ad hoc basis. Many planning regimes in developed economies have been problematic in the sense that they impede on the entrepreneurial flair of property development. For example, prior to local and central government approval, an opportunity is given to the local community to object to the development and its provisions. Such processes have been criticized as delaying the planning process. Referring to UK practice as an example, the Planning and Compulsory Purchase Act 2004, requires the local authority to consider any objection made to a development plan document and major planning applications. Such mechanisms do not exist in the Dubai planning system and development schemes are set up behind "closed doors". That does not mean that the Dubai planning system is free from community-based decision-making nor lacking any principles of sustainability – yet one must appreciate these schematic designs and ultimately their take-up in the market is much more market-driven rather than centrally controlled.

A summary of the key masterplan/development steps would be:

- Land acquisition
- Masterplan approval
- Masterplan development strategy and phasing
- Community amenities and retail strategy
- Activation of subprojects: tender/off-plan sales launch
- Design stages: schematics
- Becoming sales-ready (authorities, Dubai Land Department (DLD) pre-registration, escrow, legal, marketing and sales)
- Construction tender and build stage
- Landscaping
- Final DLD project registration and surveying
- Building completion and handover

- Building operations: OA
- Master community management

Developer regulations in Dubai

As a global comparison, in the UK, the development process is, for the most part, administered by individual local authorities in accordance with the Planning Act (s) which have been enacted in each of the devolved legislatures of England, Scotland, Wales and Northern Ireland. Local authorities have planning powers which are devolved from central government legislation and are empowered to assess developer planning submissions against local development plans and relevant national planning guidance and legislation, as well as being responsible for building control and enforcement of relevant building regulations with respect to the development and importantly, administration of any developer contributions, known as the Community Infrastructure Levy (CIL) in England and Wales, for example.

   Developers in the UK will, in addition, often be required to liaise with infrastructure authorities, including (although not limited to):

1   The local authority who will usually also be responsible for waste collection and environmental services.
2   Highways England and Network rail (and where applicable rail network operators) where development may have an impact on the strategic transport network.
3   Utilities companies, which can include: potable water supply, electrical power supply, gas supply and telecom networks, for example.

The development process in Dubai is quite similar to that described above, with obvious differences being the specific authorities that a developer must engage with in Dubai, and also in regard to the specific regulations of which a developer must be aware of and certain specific planning and approvals procedures which are unique to Dubai. Of particular importance is the role of Dubai Municipality, which is the primary authority responsible for the administration of the planning and approval process for single-plot developments and master-planned developments in Dubai. Box 10.1 provides a summary of the real estate development process in Dubai, as sourced from the DLD.

Box 10.1 The real estate development process in Dubai

The Dubai real estate development process is separated into three main stages when it comes to licensing and approval. These include (DLD, 2022):

---

1   The pre-development stage
2   The development stage
3   The post-development stage

Within each of these stages, there is also a range of developer activities that would ensure the project is delivered successfully. A summary of each of these stages and its component parts include (DLD, 2022):

Stage 1 – The pre-development stage (6 sub-stages)

- Submission of the initial approval certificate: Documents would include the completion of a commercial registration and licensing form; Emirates ID/passport number; residence visa; letter of no objection from the sponsor; no objection from the free zone; a copy of the parent's company commercial registration certification.
- Trade name reservation: Submission to Dubai Economy.
- Issuing the trade licence request: Submission to Dubai Economy with some fees also payable to DLD.
- No objection certificate (NOC) DLD approval for the trade licence. Submission of the title deed for the land under the name of the owners (same as trade licence); submission of good conduct certificate issued from Dubai Police.
- Registering the developer in the real estate developer's log. Any real estate development company to register in the real estate developers register. All approved developers have the right to register a project that they wish to sell its units off plan.
- Registration in the Oqood system course. Supported training course for developer or associated staff to learn how to use the Oqood system ready for off plan sales/registrations

Stage 2 – The development stage (10 sub-stages)

- Apply for the approval of urban projects plans/plans modifications: obtain an approval for the masterplans of major urban development projects, or an approval for the masterplan modification, according to the planning requirements adopted by the Urban Planning and Studies department (including affection plan; NOCs from relevant parties; copy of planning report approved by DM)
- Project name NOC from the master developer: approval of the project name from the master developer via a NOC process
- Project registration and masterplan submissions (ma'lem Dubai): allows for approved projects to apply for power and water supply from DEWA

- Accreditation of the Department of Tourism and Commerce: if the project includes a hotel provide a number of rooms and supporting activities
- Accreditation of Emirates Telecommunications (Etisalat/Du): assigned access to telecommunication services
- Accreditation of The General Directorate of Civil Defence (DCD): adoption of engineering plans for new buildings and installations, or upon alteration of space, or addition of construction works or change of type of works after safety requirements are met under the UAE handbook for fire and life protection.
- Accreditation of Dubai Corporation for Ambulance Services: allocation of a land plot and building of ambulance posting for each 10,000 of the population
- Accreditation of The Roads and Transports Authority (RTA): approval of the structural plan and traffic impact study for the master development project, and the signing of the partnership agreement (the developer should not start any work before finalizing this step and updating the structural plan based on the Dubai Municipality's survey system), alongside a NOC on infrastructure design
- Approval from Dubai Civil Aviation Authority (DCAA) NOC for building heights to ensure no conflicts with flight movements/paths
- Approval from Dubai Police: The developer is obligated to allocate a plot of land based on the population density in the project, an area of not less than 25,000 sq ft is required if the population density is more than 20,000; and an area of not less than 12,000 sq ft is required if the population density less than 20,000 people and the land ownership will be under Dubai Police General Headquarters.

Stage 3 – The post-development stage (2 sub-stages)

- Settlement of Escrow Account: allowing for the developer to settle the escrow account and receive deposits due on completion
- Unit Loadings in order to extract property ownership deeds for each real estate unit

Source: Summarized from the DLD (2022).

Dubai Municipality is also responsible for designating the land use and density on all land throughout the Emirate, which is documented in the Dubai Urban Master Plan. At the time of writing, the Dubai 2040 Urban Master Plan is the latest document showing the masterplan for the Emirate.

The allowable land uses and densities are documented on what are known as Affection Plans, which are documents issued by the DLD when the ownership (title) of a plot is changed (for example, through the acquisition of the plot by a new owner).

There are a number of other authorities in Dubai which are key to the development process. Each authority has individual control over the development and planning process, and the provision of infrastructure, as follows:

i   Dubai Land Department (DLD): responsible for the registration of property in the Emirate and the Real Estate Regulatory Agency (RERA) who regulate the real estate sector.
ii Roads and Transport Authority (RTA): responsible for the road and rail transit networks.
iii Dubai Municipality (Dubai Municipality): besides their planning role described above, responsible for general construction standards, waste and wastewater, irrigation and community facilities.
iv Dubai Development Authority (DDA) who since 2018 (Law No. 10 of 2018) have had a broad mandate for certain development control, masterplan and building approvals processes across a number of jurisdictions in Dubai, these typically being free zones and large master-planned communities by various developers in Dubai.
v   Department of Electricity and Water (DEWA) who have responsibility for the power supply network and the potable water supply network.

In addition, Etisalat and Du are the two telecommunication companies that provide telecoms infrastructure. Typically, in my experience, the telecommunications company selected by the developer will require that a project provides the internal infrastructure in accordance with its specifications, which may be subject to testing and approval by one of its engineers prior to the connection of the project to the external telecommunications network.

Each of the above authorities has established procedures and practices for the provision of infrastructure to a project. Put simply, these procedures allow authorities to coordinate the planning of the provision of infrastructure to projects.

Finally, it is important to recognize that in Dubai there are certain differences in regard to how development projects may take place in Dubai, to the approaches which may be familiar to developers from other geographies. These differences also have certain implications in regard to the planning process and the statutory and regulatory authorities that a developer would need to be aware of, and I have briefly described these below.

Development within a master development

In Dubai, and indeed throughout the Gulf Region, it is common for a developer to assemble very large numbers of plots, sometimes with total development areas up to 250 sq km, with a view to sub-diving that total area into a number of large plots which themselves might be sub-divided into individual plot developments.

In effect, this creates the following relationships, each with its own development procedures:

1 Master Developer: usually the initiator of the development, responsible for the original land assembly.

   The master developer will usually prepare a masterplan for the overall development, which is often comparable to a city-scale urban plan. These masterplans are usually not granular to the level of individual plots, but they will define the planned land use types and densities, forecast populations and future infrastructure requirements.

   These masterplans require approval by the relevant authorities mandated for those approvals at the time of the project. Clearly, due to the scale of these master development projects, the Dubai authorities will have a particular interest in the masterplan and forecast populations and infrastructure requirements, since the impact of the development upon the existing infrastructure network and future planning could be significant.

   Once approved by the relevant authorities, the masterplans become fixed, and any developer of land within the master development will be required to adhere to the principles set out in the masterplan.

   To this end, a master developer will usually prepare detailed development guidelines and procedures, which will be used to ensure that all developers within the master development adhere to the masterplan. This requirement will usually also be made explicit within any sales or development agreement between the master developer and any sub-developer or plot developer.

2 Sub-master Developer: a developer who acquires a sub-divided area of an overall master development, which itself comprises a number of plots.

   This sub-master development will prepare a masterplan which will be required to adhere to the master development masterplan and will customarily require the approval of the master developer before it is then submitted to the relevant Dubai approvals authorities depending upon the jurisdiction.

   Once this sub-development masterplan is approved it will then form the basis for approvals of any plot development within the sub-development and this will usually be set out in any development or sales agreement. It is customary that the sub-developer will themselves prepare development guidelines which will inform the plot development process, including controls on land use types, building typologies, heights and built areas, for example.

3 Plot Developer: a developer of a plot within a master development or a sub-master development.

   This plot developer will be required to develop the plot in accordance with any sales or development agreement, which will almost certainly require adherence to the development guidelines. Approvals for a plot development are likely to first require approval by either the sub-master developer or master developer, and subsequently the Dubai authority approvals as would customarily be required for any plot development in Dubai.

Development within a free zone

In Dubai, certain zones have been created which allow foreign ownership or long-term leases of land (where this would not be possible elsewhere in the Emirate). These zones are master planned with a strict development framework and guidelines, which are normally approved by the DDA, which at the time of writing has jurisdiction for real estate planning and development control across Freezone clusters and various master developments across Dubai.

On occasion, a dedicated development authority is established for the zone, which has certain delegated powers from Dubai Municipality and perhaps other authorities, which allows the development authority to undertake certain approvals on behalf of the Dubai authorities, as long as development takes place strictly in accordance with the masterplan as approved by those authorities.

A developer within the zone must adhere to those guidelines and may also be required to seek approvals from the Dubai authorities depending upon the obligations contained within the sales or development agreement, and indeed the scope and extent of any delegated powers which the development authority themselves may have to provide such approvals.

The development process in this scenario thus usually requires that a plot developer within the zone has strict conditions placed upon the land which they have acquired and they must therefore develop within the scope and extent of those conditions.

Development in isolation

This final scenario involves the development of a plot (or plots) of land, or indeed a masterplan, which is not governed by any development or planning controls (such as a Free Zone or Master Development Guidelines) other than those administered by the Dubai planning and statutory authorities.

This would typically comprise a development project where the developer has acquired one or more plots of land which they have the right to develop.

The developer may wish to develop a single building, in which case they would be required to follow the Dubai Municipality procedures for a building permit if the proposed building was in accordance with the land use type indicated on the Affection Plan, or alternatively, if the developer is planning to develop a number of buildings on the plot(s) they would be required to follow the DDA or Dubai Municipality masterplan submission process which are broadly similar in form and which I discuss later in the text.

Development regulatory and statutory authorities in Dubai

Dubai Municipality

The Dubai Municipality provide centralized planning and development control services for the whole of the Emirate with the exception of those jurisdictions which have been mandated to the DDA.

This includes the development and regular update of the Dubai urban masterplan (see Dubai Municipality, 2020) and administration of all planning regulations related to the development planning and approval process, as well as construction processes and building control. Importantly for prospective developers and their consultants, the Dubai Municipality also has a number of other powers regulating construction and engineering activities in Dubai. These include the regulation of the engineering and construction sector, as well as design and construction standards.

The Dubai Municipality planning procedures also integrate processes for approval of any development planning applications with other regulatory and statutory authorities through what is known as the Master Plan Approval process which includes a NOC (No Objection Certificate) process requiring individual approvals at each master planning, design and construction stage, from all of the Dubai statutory and regulatory authorities and entities. I discuss this process in detail in the following section.

Dubai Development Authority

The DDA was established in 2014 under Law No. 15 of 2014 with its mandate being revised under Law No. 10 of 2018. DDA is mandated to provide the following services across all jurisdictions, community developments and strategic projects and industries within its jurisdictions:

• Real estate planning and control
• Regulatory and licensing services
• Industry development

DDA has adopted a masterplan approval and building permit process which is very similar to that established under Dubai Municipality.

Dubai Land Department

Another important regulatory authority in Dubai is the DLD, which has an important role in the registration of land and transactions in Dubai and since 2007 has had further responsibility, in the form of the Real Estate Regulatory Agency (RERA) for the regulation of the real estate sector in Dubai (Law No. 16 of 2007).

Оставn

This is a very important role which all prospective developers in Dubai should be aware of, ensuring that sales of all real estate land and built assets meet certain requirements including (although not limited to) the licensing of companies and professionals engaged in the real estate sector, the registration of title, valuation of land and built assets, registration of sales and leasing contracts, and of particular importance in regard to the acquisition or operation of developed assets, the monitoring of owners associations.

Thus, it should be clear that whilst there may be some similarities in regard to the general principles guiding the development process in Dubai and other international locations such as the UK, that nonetheless there are also some key differences that prospective developers should be aware of. The following section aims to explain some of these differences.

Other important statutory authorities

Infrastructure in Dubai is provided by a number of statutory authorities, each with individual control over a specific utility, as follows:

i   Roads and Transport Authority (RTA);
ii Dubai Municipality (Dubai Municipality);
iii Department of Electricity and Water (DEWA).

As referenced previously, these stakeholders, alongside the telecommunication companies (Etisalat and Du) will be responsible for the provision of infrastructure plans in new development in Dubai. One might observe that there are some stakeholders, such as Dubai Municipality, who play a significant role across many of the development processes, whereas others feed into specific, more localized aspects. The approval processes across each of these authorities does support evidence that development planning in Dubai follows a standardized approach. A pertinent point to make here would be the importance of developers following the sequential order of these processes to avoid delay.

These authorities can usually only provide infrastructure after substantial lead times following the detailed definition of a particular project's requirements and after close coordination with the individual developers and master developers (Figure 10.1).

The following paragraphs provide further detail in regard to the role of each of the above authorities and the relevant processes required by each of them for the provision of infrastructure to a project.

1   *Roads and Transport Authority ("RTA")*
    The RTA is responsible for the planning, operation and maintenance of all public roads infrastructure and public transportation in Dubai. This includes the planning, design and management of the rights of way (ROW) running alongside the roads and within which all the main utilities (including potable water pipelines, irrigation water pipelines,

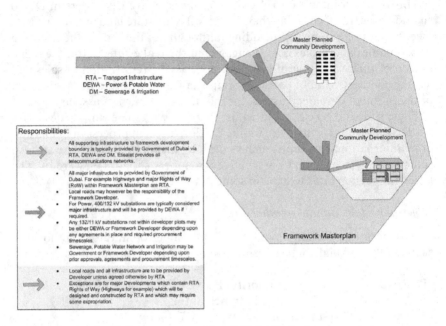

Figure 10.1 Typical infrastructure responsibilities (Dubai).
Source: Image provided by Mike Wing.

sewage lines, power cables and telecom cables) are accommodated. This is why, since its formation in 2005, the RTA has been the main coordinating authority for the planning and construction of utilities infrastructure in Dubai (prior to which it was Dubai Municipality).

Given the physical constraints limiting the width of the ROW, the planning of the utilities area within a ROW requires agreement and coordination amongst all of the authorities so as to capture as much of their future requirements as would reasonably be possible. Once the future requirements are fixed, the corridor is fixed. The opportunity for implementing further infrastructure, or of modifying the future requirements agreed upon between the authorities, is very limited once the corridor has been finalized and the road infrastructure has been completed.

In order to ensure that the width of the planned ROW will allow proper accommodation of the future utilities' infrastructure, the RTA requires that any changes to a road, or infrastructure, must go through two separate approval processes (or NOC processes): one for the design and one for construction.

These processes (presently administered by the RTA) provide an opportunity for all utilities authorities occupying the ROW under consideration, to provide their agreement with regard to any changes to the areas allocated to them for current and future utilities. Both processes must be completed for any project requiring new infrastructure to its boundary, such as power, water, telecoms, sewerage and irrigation. The construction NOC process must be completed before any construction of utilities infrastructure in the ROW can be undertaken.

The design NOC process requires the following (as a minimum) to be completed prior to the commencement of a construction NOC:

i   Approved Final Master Plan;
ii   Approval (RTA) of Traffic Impact Assessment;
iii   Approved Power Supply Master Plan;
iv   Approval (DEWA) of Load NOCs;
v   Approval (DEWA) of Water Transmission NOC;
vi   Dubai Municipality approval of Sewerage, Stormwater and TSE NOCs;
vii   Etisalat/ Du approval of Telecommunications NOCs.

Therefore, the planning of ROWs and the strategic planning for future utilities requirements is a complicated process and is largely dependent upon utilities authorities receiving accurate and up-to-date information on future development projects in order to ensure a reasonable level of certainty in regard to future planning.

The ROW planning process usually occurs in conjunction with strategic planning studies (such as region-wide transport masterplans) which determine road width requirements within ROWs in order to cater for future forecasts of traffic flows. These strategic studies are supplemented by more localized Traffic Impact Studies or Assessments ("TIA"), which deal with the specific impacts of any planned project which connects to the public highway and transport network which is maintained by RTA (Article 6 of Law No. 6 of 2006).

In order for a project to be provided with access to the external road and transport network, a TIA must first be completed and approved by the RTA. A cost-share agreement must then be signed (Law No. 5 of 2007), and the developer must agree to certain responsibilities with regard to the payment towards access to the project as well as any facilities external to the development deemed by RTA to require improvement as a result of the project.

2   *Dubai Electricity and Water Authority ("DEWA") – Provision of Power*
DEWA is the public authority responsible for, amongst other things, the provision of power in the Emirate of Dubai. The power transmission network in Dubai comprises the following key elements (as defined by DEWA's "Connection Guidelines for Distributed Renewable Resources Generation Connected to the Distribution Network" (August 2015))

i   Substations, which include the following:
  a   **400/132 kV substations**, which may be considered the primary source of power to a network. These substations step down

the main power transmitted by the power plants (400 kV) and
provide power at a very high voltage (132 kW) which can be
transmitted over relatively long distances (tens of kilometres,
for example) without significant loss of efficiency; and

b   **132/11 kV substations**, which further reduce the voltage received
from the 400/132 kV substations to 11 kV for shorter transmission
lengths (shorter lengths of cabling to the final location to
be supplied with power). These substations are often located
within, or close to the centre of development projects such as
the Project in accordance with the DEWA guidelines.

ii Transmission Network – The system belonging to DEWA which
comprises the high-voltage (>33 kV) electricity cables, lines and
electricity installations and facilities owned and/or operated by
DEWA and used to transmit electricity from a power unit to a
power substation or other electricity generation unit.

iii Distribution Network – The "medium (6.6, 11 or 33 kV) or low voltage (0.4 kV) electricity grid for supplying electricity to the end consumers". This is the network which connects directly to villas and
buildings within a development. This network typically comprises
rings of cables which connect to small substations and transformers
at the medium voltage level and provides supply to buildings at the
low voltage level.

For developers with larger projects which include power supply requirements greater than 20 MW, a developer is likely to be required to
provide its own 132/11 kV substation at the "centre of load". This has
important ramifications in regard to the preparation of cost estimates
for development and should be considered as a key cost and time risk
when preparing the development feasibility.

This also requires considerable design and planning effort, and the
development manager should recognize that DEWA will require details
of the power requirement dates, as well as the substation plot, identification of all cable corridors, ducting details and road crossing details
and the 'Total Connected Load of the Project and Expected Maximum
Demand' at a very early stage (ideally early in the design stages) order
to ensure that the substation can be constructed and connected in time
for the opening of the development project.

Indeed, DEWA suggests a timeframe of 30 months for the construction of a 132/11 kV substation after "finalization of load requirements,
substation locations, cable corridors and receiving the affection plan of
the substation in the ownership of DEWA". Thus, it should be clear to a
development manager that:

I   An assessment of power supply requirements should be made at
pre-feasibility stages

II The cost and time implications of power supply requirements are very important risk items which should be clearly understood in the development strategy

Finally, in common with the masterplan process, the development manager should understand the risks of any changes to the development once DEWA has agreed to the design submissions. Should the project change, whether as a result of design changes, scheduling changes or any other changes the developer may well be required to freeze the project, and commence a new approvals process with Dubai Municipality including obtaining new NOC from the relevant authorities.

Once a project has been constructed (which as a matter of custom and practice requires completion of all of the preliminary and design stages), and all of the internal power distribution infrastructures have been completed, it is important for a developer to understand the DEWA require that the internal power distribution network must be inspected and tested by DEWA, prior to connection to the external power network (Figure 10.2).

3   *Dubai Electricity and Water Authority ("DEWA") – Provision of Water*

The DEWA water transmission department is the public authority responsible for the provision of potable (drinking) water in the Emirate of Dubai. In order to facilitate the understanding of the processes and procedures for the provision of potable water to a new project, DEWA has issued a number of guidelines advising on the planning, design and approvals process. The guidelines are intended to help understand

DEWA substation	Project <20 MVA	Project >20 MVA
• Developer advises DEWA project requires <20 MVA (typically small project, low density or single buildings) • Developer advises DEWA project requires >20 MVA (typically higher density, mixed use, masterplanned communities)	• Shared substation located external to project, either provided by DEWA or Master Developer • Developer may be liable for portion of cost (upon agreement with Master Developer) • Internal power supply of specific project always responsibility of Developer	• Developer substation located within project • Developer bares cost for substation • Internal power supply always responsibility of Developer

Figure 10.2 Typical process flow for provision of electrical power (Dubai).
Source: Mike Wing.

DEWA requirements and facilitate the preparation of development projects masterplans and related documents.

New development projects need to be carefully studied by DEWA in order to plan to meet water demands and other system requirements, which may involve building new transmission and distribution networks or even increasing production capacity.

For example, DEWA does not supply potable water for:

i   *Construction purposes (particularly if there is no existing developed network at the project area);*
ii   *Water features (lagoons, etc.);*
iii *Irrigation/landscape purposes.*

With regard to the timing of provision of water supply to a project, DEWA also notes the following:

> Because demand and its phasing represent the most crucial element for the whole water transmission network planning process, developers are required to timely provide the following information in their submissions to DEWA:
>
> i   Reasonably projected demand figures.
> ii   Reasonable demand phasing throughout the development planning period. Each planning phase should be represented by commissioning dates rather than construction start dates. For Mega Projects, information for each phase should include the relevant small projects and their demands.
> iii Base information and calculations used to determine the water demands such as population, land use and district cooling estimates.

DEWA also emphasizes that lead times can be long: 2–3 years for distribution systems and pumping stations, 3–4 years for new transmission pipelines and at least five years for new production facilities. Furthermore, developers must provide timely information to DEWA including a masterplan for the project with population calculations, total demand phased by year, plot and demand type and a statement of availability of plots and corridors as per DEWA requirements.

Once approval has been granted by DEWA following review of the documents above, the following submissions are required:

i   Projects demands table as approved by Water Transmission Planning Department.
ii   Expected date of connection to DEWA main lines.
iii Road cross-sections and details of DEWA corridor and also details of connection with DEWA main lines.
iv   Approximate location of house connections to be shown on the layout drawings.

v Bills of Quantities for NOC with related drawings attached as per DEWA standards.

DEWA, therefore, requires quite explicit information from a developer of a project, which in my experience requires early advancement of the project's preliminary and detailed design, in order to plan for the connection of that project to the water transmission network. This is evident because the development of connection details with DEWA main lines, location of house connections and Bills of Quantities all require that the design stages are well advanced.

4   *Dubai Municipality – Infrastructure and utilities*

The Dubai Municipality is the public authority responsible for the provision of key public services such as waste collection, sewerage disposal and treatment and the management of public facilities such as parks. Relevantly, Dubai Municipality is responsible for the provision of the following utilities services:

i   Sewerage Treatment and Disposal;
ii   Storm-water Drainage;
iii Irrigation Water.

5   *Dubai Municipality and DDA (Master Plans and Building Permits)*

Dubai Municipality and DDA are also responsible for granting planning permits for new projects or changes to projects within their respective jurisdictions, as well as building permits which enable construction activities to commence. Without a building permit, developers cannot start the construction of their projects.

In order to obtain a building permit, a developer must either intend to develop a single building on a plot in accordance with an Affection Plan for that plot or where multiple buildings are to be developed, the developer must submit a proposed masterplan for its project for the relevant authority's approval (see DDA, 2020). Following approval of the masterplan, the developer must submit the detailed designs (matching the approved masterplan) for approval by Dubai Municipality. This will enable Dubai Municipality to issue the building permit which then allows construction to commence.

The approval of the masterplan is, therefore, a key stage in the overall approvals process for a development project. Furthermore, the process effectively commits a developer to a fixed masterplan once it has been approved by Dubai Municipality, since any subsequent changes to the masterplan (particularly in relation to the Gross Floor Area ("GFA") or the number and type of buildings to be constructed) will typically require a full resubmission of the revised masterplan for approval, and effectively re-start all infrastructure approvals which may have been ongoing at the time of the revision. It is possible on occasion to obtain certain dispensations from Dubai Municipality which will allow for ongoing changes to certain aspects of the masterplan; however, the opportunity to make changes is limited, and there are other planning impacts which may result from such changes which could typically delay the overall planning process including the possibility that the Traffic Impact Assessment may need to be updated, and as discussed regarding DEWA, the approvals process for connection to power would then effectively need to be restarted.

How to evaluate development in Dubai?

The previous sections have set out an outline of the property development process in Dubai. The initial parts of the chapter outlined the main stakeholders involved in property development. Of notable interest in my analysis in this chapter is the discussion relating to development risks and identifying what these are and how they are typically managed and mitigated. As with global practices major property development will start with either a decision on a development concept and/or a site acquisition. The residual valuation approach provides the main rationale on whether a development seems financially viable or not. One would often use an asking price of a site against suitable land sales and the residual land value to assess the initial viability. The next section provides an overview of the residual land valuation technique.

Residual land valuations: The basics

The sales price of completed development is often termed as the "Gross Development Value" (GDV) (calculated off the principal valuation method of the proposed asset being developed).

Less   cost of development
Less   profit allowance
Equals   the "Residual" (*the maximum sum left for purchasing the land*)

Table 10.1 summarizes where a practitioner in Dubai may collect data for these calculations.

What techniques are used in development feasibilities?

The valuation of land with development potential has been confined to the application of two core techniques: 1) the developer's budget (or residual land valuation) and 2) the discounted cash flow (DCF) approach. Both are widely used methods of evaluating viability in property development. The first technique (the developer's budget) is often performed in early stages of feasibilities to scope out whether the opportunity is viable or not. This analysis involves "balancing the budget", the value (or income) side and the cost side (including the land price), should balance. By subtracting the known cost items from the total value (or income) side of the budget, the unknown

Table 10.1 Examples of sources of advice for each variable factor

Valuation input	Source of information
Gross development value	
Sales price	DLD; Property Monitor; Data Finder (Property Finder); REIDIN
Rents	DLD Rental Index; DLD; Property Monitor; Data Finder; REIDIN; global consultants
Yields	Major property global consultants
Purchase costs	DLD
Construction costs	MEED; RICS
Professional fees	Major global cost consultants; 10% for large projects; 15% for small projects
Letting fees	5% of gross annual rent (residential and commercial)
Selling fees	2% of selling price (i.e., Net Development Value)
Finance rates	*specific to developer (typically ranges 6%–8%)*
Inflation	Trading Economics
Developer's profit margin	*specific to developer (typically ranges 20–30%)*

Source: Author's own.

cost (often the land price) can be derived (i.e., the figure that balances the budget). What is left over (i.e., the balancing sum or "residual"), therefore, represents the maximum sum the developer could offer to purchase the development site. The second technique (the DCF) is employed once the development concept and site have been suitably established, required in most cases for arranging property development finance, notably due to the ability for a DCF to be more detailed; time-sensitive; and it allows for value and cost items to be mapped out on a period-by-period basis. Both techniques overlook several factors. The sheer number of data assumptions that must be formed to undertake the appraisal, as well as the high level of interactions that take place within the dataset means the entire process is rather subjective and perhaps more worrying, prone to large variances. The technique is largely criticized both from a behavioural aspect (a practitioner could manipulate inputs to arrive at the desired answer) as well as technical (small changes (or errors) in the calculations lead to vastly different results). The covariance between the techniques' many data assumptions also means that the magnitude of error, when errors do occur, is both professionally and financially crippling.

Even when a developer is using the method in a "professional" manner, the figures used are estimates or guesses. Therefore, a large limitation in development feasibility relates to the uncertainty surrounding the figures used in the appraisal. Even when a developer makes every effort to obtain the "correct" figures, these are still "guesses", and running feasibilities in opaque markets, makes this even more challenging. Consequently, arriving at the "right" answer is problematic. Thorough market research and running sensitivity testing is one way of addressing these deficiencies however that requires the right data to be available in the first instance. It should also be added that using a cash-flow approach does not make the values of the variables used any more "certain" or accurate than they would be in a conventional budget appraisal (i.e., the issues of uncertainty and risk are still present).

During property booms (2002–2008), overlooked elements or erroneous spreadsheet models may have not seen a project fail as any shortfall in the required developer's profit was matched by rising rents or sales values (income). However, when there is a crisis or recession in the property market the failings of feasibilities come to light. Post-Global Financial Crisis (GFC), developers in Dubai were more careful to evaluate the land bid and development concepts through the application of development appraisals and feasibilities – lessons had clearly been learnt.

A worked example

Before we look at a detailed example, let's first address a specific issue with development feasibilities, the geared relationship between cost and value magnifies the effect of any errors or loss. A known issue with all development appraisals is that the relationship between the residual outcome is highly geared and very sensitive to the assumed inputs (Havard and Waters 2013). This is a crucial element to understand and can be examined with some simple examples. Let us assume that an initial appraisal produces the following broad figures:

> Value on Completion AED100,000,000
> Development Costs (inc. interest) (AED60,000,000)
> Land Cost (inc. holding costs and interest) (AED23,000,000)
> Development Profit AED17,000,000
> This is a profit of 20.48% on costs (17,000,000/83,000,000).

If, between doing the appraisal and the development being completed, values fall 5%, the following happens to the development profitability:

> Value on Completion AED95,000,000
> Development Costs (inc. interest) (AED60,000,000)
> Land Cost (inc. holding costs and interest) (AED23,000,000)
> Development Profit AED12,000,000

Although this is still a 14.45% profit on cost, it can be seen that the 5% drop in values, something which can easily happen over the period from inception to completion of a development, has been magnified into a 30% drop in profitability. However, real estate markets tend not to move one variable at a time. If we put this into the context of an investment valuation whereby the resultant value is derived from rent and yield multiples, the derived value on completion has a large scope of much larger variance. For instance, in the above example, initially 100,000,000 could be derived from an annual rent of 10,000,000 and an assumed investment yield of 10% (i.e., 10,000,000/0.1 = 100,000,000). Then if rents fall 5% (to 9,500,000), yields in a downturn are also typically higher (let us say to 11%), the resultant value on completion is now 86,363,636 (9,500,000/0.11).

This by itself is worrying enough, but often deterioration in values is accompanied by an increase in the length of time to let or sell the scheme. This increase in time increases costs, essentially due to a rise in interest charges.

If we look at the drop in value combined with a 10% increase in costs produces the following effect on profitability:

Value on Completion AED86,363,636
Development Costs (inc. interest) (AED66,000,000)
Land Cost (inc. holding costs and interest) (AED25,300,000)
Development Profit (Loss) (-AED4,936,364)

The scheme makes a large loss.

The lessons from this are threefold (Havard and Waters, 2013):

1   First, the highly geared/sensitive nature of development means that only small market movements can impact viability. This is one of the reasons why development is a high-risk activity and also why repeated appraisals and research throughout the period running up to the development start are essential. These issues are exacerbated further in opaque markets, where obtaining data itself can contribute to a large variance

2   The second lesson is the impact of variance/error is not technical but is an error in forecasting. It underscores the need for the inputs to the appraisal to be as accurate and carefully considered as possible. Any assumptions should be validated and come from reliable sources. To this point, any development appraisal must consider appropriate risk and sensitivity testing to show the client the impact of a range of changes in data on profitability/land value bids.

3   Finally, it underscores the need to have a reliable model or framework for constructing the appraisals. One of the core reasons behind the presence of errors in conventional spreadsheets has been the lack of monitoring and the ability to track the changes made between users. Of course, not all spreadsheet errors are on the danger-critical scale, but given the magnitude of the monetary sums being invested, even the presence of fractional errors – could potentially be disastrous for an acquisition or disposal case in real estate. Client pressure on the appraisal can also further extend any of the above-mentioned appraisal issues.

Standard software packages, such as Estate Master DF, bring more reliability to the feasibility process, enabling the use of consistent interfaces for

cashflow forecasting that does not expose the process to error. Formulae between cells are locked, cannot be overwritten or accidentally deleted. Some
organizations with a limited budget, or lack of appreciation to the complexity involved in development feasibilities may, and indeed do, settle for an
in-house manual spreadsheet to serve these functions. Calculations and cell
linkages are typically built with a current development project in mind, let's
stay, a high-rise residential development. These same functionalities cannot be transferred over to the newly-instructed shopping mall development.
Equally realistic would be the outcome that the employee, who has spent
many a month writing the functions in the spreadsheet, is soon offered a
higher-paid position with the firm's competitor. Whether they take the model
with them or not, it would be a laborious, if not an impossible task, to dissect
the algorithms and formulae expressed in the model – it just is not a transparent way of doing business!! Banks, joint-venture partners and government
authorities also favour cash-flow appraisals undertaken on a standardized
platform, such as Estate Master, largely due to the fact that the functions
within the software are "fixed" and less prone to errors or manipulations.

Case study: Prime residential plot: for sale (DIFC, Downtown Dubai)

Now let's look at a real-world example. The case study below is drawn
on a development opportunity in Dubai International Financial Centre (DIFC). The details of the available land plot and its surrounding
land use information is shown below:

- Large freehold plot (198,000 sq ft)
- Maximum permissible GFA 1,375,000 sq ft

The plot is located with DIFC, a major financial hub (110 ha), offering
100% corporate ownership. The DIFC operates under its own legal
jurisdiction (English common law). DIFC is largely a commercial and
hospitality district and unlike Downtown Dubai, there is not a large
supply of premium residential development.

If the development opportunity deems residential as the most profitable use (assumed in this case), then a number of key comparable projects
are identified in the table below (see Table 10.2). From an analysis of the
local market, comparable high-end residential developments include:

1 Index Tower
2 Burj Daman
3 Sky Gardens
4 Limestone House
5 Central Park (excluded in this analysis)

A shortlist of appropriate comparables is then provided in Table 10.2.

Table 10.2 Key residential tower comparables

Project	Location	Units	Type	Completed	Price per sq ft	Occupancy (%)
Sky Gardens	Central DIFC	575	Residential	2008	1,650	90
Burj Daman	Al Mustaqbal St.	275	Serviced Apt	2013	1,500	88
Limestone House	Central DIFC	124	Serviced Apt	2011	1,500	98
Index	Central DIFC	520	Residential	2011	2,000	93

Source: Supplied by Andrew Love.

Development appraisal and assumptions

The appraisal below shows that the developer (based on the residential schematic assumed) is seeking to make a profit of 30%. A profit-on-cost reference is largely a better way of analysing the profit for the developer's risk on capital invested in the project. However, given the land value is part of the development cost and is unknown, profit on value can be used as a suitable estimate. With this target return in mind, a rational developer would be willing to pay AED260m for the land purchase (see Table 10.3).

In this scenario, a DCF cash-flow timeline can be constructed of the assumed timescales as shown in Figure 10.3.

A DCF would typically be required when looking to secure finance for the scheme. No allowances have been made for phasing the sales (or pre-sales) that might make the proposition more attractive, and bring forward some of the income in the project. At present, all income is only received in the last quarter.

Prudently within the analysis, a practitioner would cross reference the output of their appraisal figure with surrounding land prices to sense check whether the appraisal is in line with recent comparable land sales. DIFC is an established area with very few large land plots available, therefore, suggesting land prices might begin to increase, particularly for those plots located on the main transport roads. Some notable land price comparables include:

- Business Bay is located in close proximity yet a less desirable area within Dubai (achieved land sales in the region of AED300–AED350 psf GFA).
- Downtown Dubai is approximately AED375 psf GFA

Table 10.3 Basic residual development appraisal for DIFC plot

Asset	GFA sf	Efficiency	Sellable area	Net sale AED/sf	Sales value
Residential (sales)	750,000	85%	637,500	2,000	1,275,000,000
Branded residences (sales)	625,000	85%	531,250	2,500	1,328,125,000
GDV	1,375,000		1,168,750		2,603,125,000
NDV (after 5% transfer fees)					2,479,166,667
Land purchase price					(unknown)
Costs					
Land acquisition	5%				
Site surveys					(500,000)
Construction costs (CC)					
GFA of 1,375,000	700				(962,500,000)
Professional fees (PF)	10% of CC				(96,250,000)
Contingency	5% CC/PF				(52,937,500)
Project marketing	1.5%	GDV			(39,046,875)
Agency (sales)	5%	NDV			(123,958,333)
Finance cost	6%				
Interest from design stage	assumed as 3 months on site survey costs and professionals fees				(1,419,695)
Interest from construction	assumed as 15 months (30-month build phase/2) on build cost, contingency and project marketing				(79,670,805)
Interest from sale holding period	assumed as 3 months on all costs excluding site surveys/agency fees				(16,885,699)
Total project cost					1,373,168,907
Developer profit (30% GDV)					(780,937,500)
Residual land (future)					325,060,260
Present land value (*)				0.8396	272,926,862
Land acquisition costs	5%				(12,996,517)
Residual land value ("bid")					259,930,345

Source: data calculations and assumptions by the author., * total development period is 36 months, with assumed discount rate 6%.

The appraisal on this plot is showing an AED189 psf GFA, therefore, suggesting that although the feasibility is attractive to the developer (30% profit) it is unlikely to win the competitive tender as the developer's bid is somewhat below what land prices have sold for in the vicinity. A prudent developer might re-evaluate the income and cost assumptions. It may also suggest the schematic needs to be higher value, perhaps with more branded residences and include some commercial assets. A developer when using a feasibility must also bear in mind the likely land sale price from the comparable evidence.

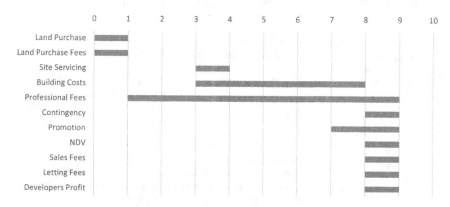

Figure 10.3 Example Gantt chart to illustrate the importance of time value of money in the RLV method.

International comparisons with Dubai

Having examined the property development process in Dubai and the range of techniques used in development feasibilities, the attention of the discussion now moves to the developer's formal decision-making.

Dubai presents itself as a fascinating case study to compare and contrast development feasibility processes given the sheer scale of mega projects and large master communities, which in themselves go far beyond the typical time scale suited to these techniques (>10 years). The mixed-use nature of many of the developments in Dubai present challenges in terms of anticipated future values; necessary infrastructure and a fast-changing development landscape that may mean what was in demand at the inception stage has become plentiful by completion.

Pre-2008 developers largely ignored feasibility analysis and land-buying speculation was rife. When these land parcels were then re-valued post-2008, owners were somewhat disappointed. Therefore, the downcycle during GFC exposed the paucity of analysis of development feasibility, most prone to large masterplan developments. Many of the earlier development projects that were not underpinned by a rigorous analysis never came to fruition; were delayed or now are thankfully being rethought based on a more analytical approach. Table 10.4 explores how development strategies in Dubai might differ somewhat from other international comparisons.

Table 10.4 highlights the challenges that property developers might face in Dubai compared to other international markets. The factors highlighted point to an increased level of "uncertainty" rather than "risk" in property development, the former being anything that is not known at the time of making a decision at the feasibility stage. Risk on the other hand can be more adequately measured and alleviated during the process. Let us examine these stages and observations in more detail:

Table 10.4 Perspectives on key stages of property development: Dubai vs global

	Dubai	*Global*
Market maturity and transparency	Less transparent (improving)	Mature, relatively stable and transparent markets
Ownership rights	Within selected area	Strong ownership rights
Planning	Power and freedom of developer(s) with development process requirements	Local government regulation/limited power for developer
Development funding	Developers often funding with 100% equity; individual investors fund development via long-term payment plans. Some JV examples	Developers typically use 30% equity, capital raising prior to development. JV and other options
Finance	Higher LTV (20%) with higher bank margins (risk premiums)	Lower LTV with low interest rates

Source: Author's own/Erik Volkers.

- **Market maturity and transparency:** If the market is more volatile and less transparent then the outcome of a development appraisal is less likely to match the reality of the market when it is completed. Therefore, the developer has two options – they accept the heightened risk (and speculation of the development) and increase the allocated % of profit/risk contingency to take out of the project or they pre-lease/pre-sell as much of the project as possible (and lose out on any potential future upside in future sales rates). Over recent years an increasing number of reliable data sets have become available that measure key parts of the development appraisal, these will continue to expand and alleviate this historical challenge.
- **Planning powers:** Liberal planning powers mean that there are fewer restrictions upon new supply coming into the market, making property development in Dubai much more speculative than in markets that have more restrictive planning laws. The outcome is that foresight on the estimated supply is challenging for a developer at the conceptual stage, and cannot be a managed risk throughout the development process. Information on new supply however has improved in recent years and this does serve developers well in better understanding levels of potential future competition.
- **Development finance and funding:** With limited finance options, high borrowing rates, off-plan sales become a source of funding for developers (Box 10.2 expands on this further). It is also common for some developers to fund their schemes with 100% equity.

Box 10.2 Cycles in the sand

There is nothing more pronounced in Dubai than the development of some of the world's most iconic developments, such as Burj Khalifa and Palm Jumeirah in Dubai. Yet within short distances from these high-profile projects, other developments have never completed. What made the Palm a global mega-success while others never took off?

It would be apparent that bad timing might explain the major reason. Time is one element of the property development process that is an accepted risk but has important considerations beyond simply being able to plan and manage a build-out period. Property development is a dynamic process and is vulnerable to economic and financial changes This was never as apparent as the GFC in 2008, where the prices of properties being sold on some developments fell by as much as 50%. Well, what impact do these types of market shifts have on a development feasibility? Let us take a simple scenario based on a theoretical schematic:

GDV (Sales income): 10,000,000
Development costs (inc. land): 8,000,000
Profit: +2,000,000 (20% GDC)

If the sales income (GDV) falls by 50%, then the developer faces a development loss of -3,000,000. Even if 60% of sales were pre-sold at the original prices as of the feasibility, the developer would still only break even. In this case, as the development has minimal development on it (major infrastructure had not been expended), then the developer is likely to respond by cancelling/re-evaluating the development project, as continuing to develop would carry the significant and uncertain future development risk.

The element of time is also a fundamental factor, especially for projects that go beyond ten years. This creates heightened uncertainty about what market conditions will exist on completion. Within a scheme, there is also complexities around phasing to release buildings in the "right" sequence so as to avoid segregation or price competition within the scheme itself. Box 10.2 provides a short analysis of development risk in Dubai and highlights the contrast between two very similar development concepts, yet two very different development outcomes. The analysis is a means of highlighting the importance of timing and cyclical risk.

Box 10.2 highlights only one aspect of the property development decision. Developer's attitudes to risk pre-2008 were built very much of a city expansion plan "build it, they will come", many of which did not carry out a development feasibility. At the time, there was very much a speculative view of land buying and subsequent development. Those late to the table (in the run-up to the impact of the GFC), found their projects abandoned, as developers could not financially bear the cost of a significant drop in property values. Adding further to the issue, was that of information uncertainty. Even if a developer was to sit down a run a number of sophisticated feasibility models, where would they look to get reliable data? There was simply a lack of transparency pre-2008 (as discussed in Chapter 8). Until quite recently, developers would be often second-guessing the value the proposed developments would command some 30–40 months to project completion (especially as a supply pipeline over a five-year period was often opaque and based on an assumption of a developer realization rate between launches and completions). Developers would, therefore, be going in somewhat blind to the estimated income they might receive when the project is complete. This uncertainty was managed by taking a significant number of pre-sales to manage the carrying risk of end-values. This then meant individual end-users and property investors became development funders. This at the time presented a further risk for the project, as defaulting from payment plans by a small proportion of unsecured borrowers may put the whole development under jeopardy. This also exposed the majority of other investors to significant unsecured purchase risk. Using the analogy of single stock investing, an investor was putting their finances to support the operations of a single developer, many of which at the time did not have a development track record. If the developer goes into liquidation, very much the same as when a company declares bankruptcy, that investor potentially can lose all their investment. In a pre-2008 scenario, many individual investors were last in line as creditors from a case of liquidation. Pre-2008, there were no escrow laws in place in Dubai meaning that anyone investing had limited protection if the project halted or worse case never got developed. The Escrow laws of 2008 meant that nowadays funds are held in a deposit account managed by RERA, and these funds from off-plan sales, are only released to the developer at key construction milestones.

Key issues affecting development feasibility in Dubai

As explained in the previous sections, there are a number of defined approvals processes in Dubai which are administered by certain authorities and which should be considered at the early stages of project feasibility in order to determine the following factors.

Project timescale

Project timescale has a considerable impact on the project feasibility, in particular with regard to achieving the projected sales or leasing target dates, and thus project revenues.

The following issues should be given ample consideration at the pre-feasibility and pre-concept stages of the project by anyone involved in the project planning and feasibility:

1 A realistic timescale for masterplan approval by the relevant authority.
2 The building permit process, and other construction permit timescales.
3 General authority liaison requirements, as considerable day-to-day liaison is typically required whether through the eNOC and submissions process, or in response to authority requirements or inspections as they may be advised throughout the project lifecycle.
4 Requirements to complete certain studies which are dependencies in regard to the approvals processes, which at the time of writing typically include:

   a   TIA of any RTA contribution requirements
   b   Environmental Impact Studies
   c   Utilities demand analysis, including, for example, the DEWA Power Supply Master Plan and submission process

5 The impact of any changes to the project which may affect the status of prior approvals. For example, any change to the approved masterplan can require the project completely restart all of the approvals processes, and any current or outstanding approvals may be placed on hold until the revised masterplan is approved. This issue becomes even more significant in the event that the change in masterplan also requires a revised traffic impact study or revised infrastructure requirements such as power supply.
6 Where a project is of a large scale it may require its own dedicated infrastructure, as explained earlier. For example, this could include a dedicated 132/11 kV substation within a project, which may be both a cost item for the substation and a time item in regard to the typical procurement timescale for such an item. A 132/11 kV substation may take 36 months or more from the time the requirements are agreed upon with DEWA. Other key infrastructure items which should be considered include sewage treatment where no connection is available to an external network (or during the early stages of the project where the population does not produce enough sewage output for a local network to operate efficiently) which also impacts upon the planning of landscape as the sewage treatment process may have been relied upon to provide Treated Sewage Effluent (TSE) for the irrigation of the landscape.
7 Registration of the project with RERA and any other RERA processes which must be completed before sales can be progressed.
8 The issuance of Title Deeds for plots and RERA inspections of any premises or unit which allows the sales and leasing process to be finalized.

Project costs and other risk items

Furthermore, anyone involved in the feasibility of a development project in Dubai should also be cognizant of cost requirements which may affect the feasibility, which may include (although are not limited to):

1   RTA contributions for roadworks as a result of the impact of the project upon the surrounding road network.
2   Power supply costs, particularly where a dedicated substation is required, which at the time of writing can be as much as AED120 million.
3   The cost impact of project delay due to the approvals process, authority inspections or unplanned change in the project.
4   Finally, the issue of infrastructure timing and availability, as well as where the responsibility lies for the provision of certain internal or external infrastructure, is a key issue. Many real estate disputes in Dubai occur precisely because this issue has not been understood either in the development agreements between developers and master developers, or because the project feasibility has not allowed adequate provision for time and cost impacts associated with infrastructure, as well as any temporary infrastructure requirements which may be needed during construction, or in the event of any delay in achieving a connection to certain external infrastructure at the time of completion or operation of the project.

One of the most significant risks in the management of development in Dubai is the infrastructure required. The apportionment of infrastructure costs to the individual stages of the feasibility is complex and cost-dynamic. The accountability of who pays for these costs and what surrounding schemes may emerge in the future to push these costs even higher is somewhat of an unknown. In other markets, like the US or UK, a Community Infrastructure Levy (CIL) apportions costs against the scale of the development. In Dubai, this infrastructure costing is not as transparent and therefore it becomes harder to factor these costs (with certainty) into an appraisal pre-development. Any uncertainty on future costs can make the feasibility process more opaque.

Conclusions

This chapter has provided a detailed overview of the development processes in Dubai. From this analysis, a thorough number of the steps that need to be taken has been provided, allowing an extended discussion on the management of risk throughout property development in the context of Dubai. There are a variety of views on, and descriptions of, the development process. It is difficult to capture this process as a linear sequence of stages as each development is different and stages can often overlap.

Development tends to be a high-risk activity because of the long timespan involved, often three to five years, and a large number of determining factors which can move the variable planned out at the feasibility stage, in an adverse or favourable direction (e.g., rental levels, yields, building costs, finance charges). It is this complex and changing context that greatly impacts the development appraisal process. I discussed in Chapters 8 and 9 about the historically opaque nature of property information in Dubai, from which one would begin to appreciate the challenges of making forward-looking assumptions on development when data were opaque. As data transparency improved and the market matures it can be expected that the variance in these assumptions does reduce somewhat. Risk management processes and scenario testing in property development decision-making can be built off a more reliable starting point (than perhaps was the case pre-GFC).

The latter part of the chapter explained the basic premises of a development appraisal that utilizes the residual method (and subsequently a DCF). In the earlier years (2002–2009) many developments were undertaken on a "build it and they will come" mentality. At the onset of the financial crisis and beyond, as many land acquisitions were re-appraised, developers began to wake up to the need for a more rigorous analysis of land acquisitions. These are now broadly in line with international expectations and utilize common methodologies seen in many other global markets. How developers in Dubai approach several factors within the development appraisal, however, could be different as I have shown in the international comparisons made. As a round-up to the chapter, I examined the key development risks based on the earlier discussions on data, market maturity and the development planning processes in Dubai.

References

DDA Master Planning Guidelines (2020) https://www.dubaiapprovals.com/blog/master-plan-submission-&-approval-guidelines-of-dda.

DLD (2022) Real Estate Development Cycle. https://dubailand.gov.ae/en/open-data/developer-book/#/ (Accessed January 2022).

Dubai Municipality (2020) https://www.dm.gov.ae/wp-content/uploads/2020/08/Dubai-2020-Master-Plan-ENG.pdf.

Havard, T., and Waters, M. (2013) Evaluating Real Estate Development Feasibility Software Options, Estate Master White Paper Series. https://www.estatemaster.com/docs/default-source/whitepapers/spreadsheet-error-professional-practice-guidleine.pdf?sfvrsn=0.

Part D
Future directions

11 Sustainability

This chapter begins by setting out the global challenges for the real estate profession and through insightful discussions, point to the ever-increasing evidence that showcases Dubai as an innovator in sustainable real estate practices, dispelling the myth that sustainability is not an option in Dubai. Earlier chapters in this book pointed to the opportunity for Dubai to build up an ESG-compliant stock, that would bring new inward institutional investment. This chapter showcases a range of initiatives that have helped Dubai progress extensively in this crucial part of the profession. It provides two highlighted examples of best practice, Dubai Sustainable City and District 2020 (Expo City) as exemplar cases. The chapter concludes by looking to the future and evaluating what might be for ESG in Dubai.

The real estate sector faced significant disruption resulting from challenging dynamics created by the COVID-19 pandemic. At the same time, the significance of environmental social and governance (ESG) in property rose to the top of the agenda. Dubai has been playing its part with the increasing number of green-certified buildings and green building codes introduced. For example, Al'Safat is a green code, used to improve energy performance in new buildings. The UAE is leading the Middle East and is home to approximately two-thirds of the region's green-certified office space. Similarly, in 2019, Dubai at the city level was awarded a Leadership in Energy and Environmental Design (LEED) "platinum rating" – the first city in the Middle East and North Africa region to have that accolade. However, Dubai is not immune to the same global challenges that persist in this field of real estate. The first section of the chapter looks to examine the major global challenges facing sustainability in our sector.

What are the global challenges?

The World Economic Forum (WEF) each year produce an annual risks report. In recent years, this report has highlighted the significant climatic risks that are now present-day challenges. The report measures a wide category of global risks that are presented two-dimensionally as likelihood and impact. Over the next ten years, "climate action failure", "extreme weather",

DOI: 10.1201/9781003186908-15

and "biodiversity loss" rank as the top three most severe likely and impactful risks (WEF, 2022).

In the real estate industry, we need to be doing more to provide lasting solutions to these most likely and impactful risks. One key challenge has been an inability to agree on what exactly sustainability is; how it should be measured and whether it provides a sufficient premium to warrant the additional financial capital. While real estate development is responsible for a significant amount of global carbon emission, often the best we can aim for is relative sustainability, choosing options that are measured and more considered around environmental; economic and social goals should be a developer's priority. Thinking holistically is complex and challenging; qualitative measures are hard to incorporate into decision making strategies, but sustainability is now at the forefront of policy for many global businesses.

Historically, the attention of the solution has been on the build or building design, yet more recent evidence backs up the need to also examine building occupancy. Global building certifications can be criticized as not being the enablers of sustainability we expect. There are apparent disconnects that exist between the building design and the occupier or user of the space, that further exacerbates progress to making an impactful difference in the ways in which the built environment is used. Therefore reducing the environmental impact is not always as fulfilled as the green certifications may suggest, with one UK study finding that on average highly rated green commercial buildings can use up to four times more energy during occupancy than planned at the design stage (Innovate UK, 2016). For example, there is a wealth of literature and good examples of sustainable buildings and architecture but discussion of how well they operate is less well documented. This suggests that more needs to be done to bridge the operational performance gap so that we start to see the green ratings of buildings lead to a meaningful reduction in energy costs and carbon emissions.

Key regional challenges to sustainability, have typically been:

- Lack of stakeholder engagement – Occupiers, investors and developers have worked according to siloed needs and requirements and often lacked consultation during the development process. A sustainable transition needs to involve more engagement at both institutional and social levels.
- Higher costs – Continued misconception that sustainability efforts come at an increased cost. Large knowledge gap and a lack of empirical evidence in the market to support a business case.
- Traditional development business models – Lack of incentives to go beyond minimum regulatory requirements. Developers implement the most cost-effective building technologies to take profit from a project rather than suffer the cost of efficiency during the longer-term occupational phase.

By 2030, approximately 60% of the world's population will live in cities, according to the UN's World Cities Report 2020. Urban centres, like Dubai, play a leading role in the quest for a net-zero future. The development of real estate via the construction process accounts for 50% of raw materials global consumption and the built environment accounts for almost 40% of carbon emissions globally. Approximately one-third of these emissions occur throughout the occupancy and operations of real estate, where the remaining balance is attributed to embodied carbon and the construction process (UNEP, 2021). Therefore, the process of new development needs a radical rethink. While we know the size of our urban population is growing exponentially, are we at full efficiency within our existing stock before we pioneer new development? History suggests we are not. There is a clear environmental case for businesses to be using their commercial spaces much more efficiently before new development is permitted to take place. Furthermore, the existing stock that often falls short of new environmental performance codes, is at greater risk of becoming obsolete. A recent study reported that 45% of global commercial portfolio values could be lost to ESG market shifts and climatic risks, a sum estimated to be US$35 trillion, equivalent to the GDP of the European Union. The financial case of "doing nothing" is now very compelling.

The Middle East is set to see US$2.5 trillion worth of real estate developed by 2050 (Savills, 2021). With construction and real estate activity contributing significantly to carbon emissions globally, how will the region be able to cope with the increasing pressures of large demographic change and intense urbanization? To help answer this, the following discussion moves on to examine the evolution of sustainability globally to examine where frictions have existed in the profession and to shed light on what developments have taken place to support a more sustainable path for real estate.

The evolution of sustainability in real estate

Over 35 years ago, global governments and politicians gave us a normative vision of what sustainability is or should be. This centred on meeting the needs of the present day whilst preserving our natural resources for future generations (WCED "Our Common Future/Brundtland Report", 1987). This picked up on a disparity between economic growth of developed vs developing economies if we were to follow this definition in the strictest sense. In more recent times, we have seen waves of international agreements and coalitions that set targets for national governments. For example, the Paris agreement in 2016, with an aim of decreasing global warming, built off a 20/20/20 strategy, that being a pledge for a 20% reduction in carbon emissions; a 20% increase in the use of renewable energies and a 20% increase in energy efficiencies. Sixty percent of the global emissions savings must come from buildings. In 2019, national governments around the world began to commit to a net carbon zero by 2050 plan, built around a premise of carbon storage and off-setting. In the UAE and Dubai, similar pledges have been made.

In the last 15–20 years, whilst sustainability has been on the industry agenda, it has suffered from temporal shifts in political priorities, none more so that the management out of the global financial crisis, which saw sustainability lose some focus within real estate use and development in particular. Businesses wanted and needed to keep the bottom line of their balance sheets looking as profitable as they could be, so that might have meant occupying cheaper buildings or not investing in new greener technologies. Prior to this, the key drivers to introducing greener more sustainable buildings were through more stringent building codes as developers often questioned the marketability of green buildings in the earlier years. Developers who did seek competitive advantage of building green were often found to merely be playing "lip service" to the rhetoric of sustainability (Dixon et al., 2006). Key debates existed on the mixed views as to whether green buildings were in demand or financially desirable products. Cadman (2000) captured this in his vicious circle of blame summation of how the real estate market was essentially passing responsibility from occupier to investor; from investor to developer; and from developer to regulator as to why not as many sustainable buildings existed. Table 11.1 presents a range of 'market failures' that have impeded the development of green buildings within a global city. These factors represent areas that need 'breaking' in order for more adoption of sustainable real estate assets.

There has been an inherent implied issue with the provision of green buildings costing more to develop and payback periods are beyond the typical investor holding period. The challenge for change is exemplified by the fact that the existing established (non-green) office norms were still attractive to occupiers and investors.

Beyond legislative push factors (for example, building codes and grants), for private businesses sustainability mostly needs to satisfy the economic principles of the 'profit-maximising rule' and be financially feasible. Since then, there has been a tremendous effort from the real estate research community to de-myth these views. A multitude of academic studies have

Table 11.1 Real estate and sustainability – the "breaking" of market failures

Market failure	Description of market failure
Information	• Perception of "green value" is only anecdotal • Payback periods are too long and postpone action
Operational	• Focus on paying more for green buildings at the point of transaction rather than understanding total occupancy costs
Institutional	• High-risk premia attached to sustainability as a new property form • Focus on "market premiums" for sustainability rather than the threat of a higher rate of depreciation/obsolescence for inaction
Governance	• Lack of regulation of higher compliance and fiscal incentives to encourage the development of sustainable buildings

Source: Author's own.

provided evidence to suggest that green buildings are credible and the benefits include:

- Lowering energy and wider operating costs
- Improved workspace productivity from wellness attributes in green buildings
- Increasing Corporate Social Responsibility (CSR) and Environmental Social Governance (ESG) driving demand for businesses to occupy green buildings

The initial challenge has been moving away from quantifying the benefits of green buildings simply as a function of lower energy costs. Yet the benefits of green buildings should be embedded deeper into the fundamentals of real estate. Leskinen et al. (2020) provide a comprehensive overview of the impact of green buildings on the key parameters of real estate assets.

This analysis shows that the real estate investment criterion is leading the push to green buildings both from an occupier and investor perspective. The study found that the traditionally perceived benefit of lower operating costs in green buildings is inconclusive, yet the value drives of higher rents and lower yields, and subsequently higher sales prices as significant. This type of financial analysis shows that ESG credentials are going to play a much more meaningful role in the future of real estate asset management.

Another component "breaking the circle" are the enablers of change in the industry that are facilitating better decision-making and at the forefront of these would be valuers and strategic consultants who are openly encouraging clients to think about green initiatives in their acquisitions, new developments and workplace strategies (Lorenz, 2008). Large financial decisions are not made on anecdotal evidence, so key information about the potential for green premiums from professional bodies like the RICS and its members will help create change. Waiting for "green obsolescence" to take place is not a desirable outcome, as a large proportion of impacted stock will be the existing older buildings, which if demolished would result in a much larger impact of carbon emissions than that from proactive green maintenance. There are examples now where asset managers in Dubai are retrofitting the older stock in light of these upcoming challenges (for example, Al Thuraya Tower, Dubai Media City is being refurbished to improve its asset performance (Savills, 2021)).

Whilst the financial drivers are becoming more apparent, legislation is still playing a part in driving demand for green buildings. A step change in the premise that consumers would pay more for green buildings in Europe came about with the introduction of Energy Performance Certificates (EPC) that were legislated in 2007 under EU Directive. This law meant that any building to be rented or sold must have been energy-rated. This gave occupiers and investors a standardized industry to benchmark their decision-making when it came to buying or occupying sustainable buildings. Early studies (Waters, 2006, 2007) found these measures being used to yield a "brown discount" over a "green premium", however, businesses and institutional investors are now wide awake to the threat of sustainability left unchecked. Furthermore, any building with a lower rating (E, F or G) could not be leased or sold without energy improvements being made to bring them into the higher acceptable ratings (A to D). This represented as much as 20% of the commercial stock in the case of the UK commercial sector. Similar policy of energy labelling on renting and selling properties in Dubai might enable more change and wider development, retro-fitting and occupation of green buildings in the future. The recent announcement in 2022 that the Dubai rental index will be more closely linked to the building classification survey and rating performance may breed similar environmental regulation to impact the ESG decision-making of consumers, investors and future development.

Nowadays, there has been a greater emphasis on the occupier as part of the solution and incentivizing behaviours of developers, investors and occupiers for mutual benefit. An example of this would be the prevalence of more green leases or MoUs as these operate as an institutional mechanism that closes the gap between building ratings/design and occupational savings (which historically has meant simply leasing a green building isn't enough). In Dubai, Majid Alfuttaim (MAF) has implemented a range of these initiatives. In District 2020, tenants will be bound more by the operational efficiencies in the buildings.

The motives for occupiers to turn to sustainable buildings are clearly separated into the key drivers of: (a) reduced operational costs (choosing a green building to lower operational costs), (b)corporate governance or corporate social responsibility (choosing a green building for well-being and talent retention) and (c)branding and shareholder retention (choosing a green building as investors seek ESG goals). The main driver has always been assumed to focus on reducing the operational cost. Yet, global businesses on average spend less than 1% of their workplace costs on energy, compared to 10% on rent and 85% on employee salaries and benefits (cited in Livingstone and Ferm, 2017). Therefore, there is potentially a greater saving to be made around employee productivity, talent retention and well-being. Of course, sustainability means and matters to different business sectors to varying degrees. Competitive advantage, an influential factor in driving sustainability was not seen as significant ten years ago, but today is a standout reason.

How are developers impacting sustainable development in Dubai?

The key challenge has been how are developers addressing sustainability effectively and perhaps the simplest way of analysing developers' response to sustainability is to examine the main decision-making criteria in real estate development. As discussed in the previous chapter, there are four main decisions impacting value creation in the development process. These include:

1   Gross Development Value (or income)
2   Land value
3   Design and construction costs
4   Profit

Within each of these main stages, a developer will evaluate both risk and uncertainty. In the context of this chapter, the opportunities for most financial cost saving from a focus on sustainability are in the third stage (design and cost). That might mean the opportunity to use more environmentally sound building materials with the rising cost of more traditional factors of production. The income side of the development equation, despite academic research (as referenced above), is still somewhat unknown. How does a developer at the feasibility stage factor in a hypothetical price premium if the market is not showing evidence of higher prices? It presents an added risk for a developer to take. Developers might also add a fifth main driving factor to their list, "brand value", which encompasses corporate social governance and now ESG. For developers it might mean there is a new business case opportunity to attract new capital that is now seeking the highest levels of environmental performance. Energy metrics and labelling has made this a bigger part of the decision-making process, particularly more so in the commercial occupier market (CSR and branding). Though nowadays residential eco-premiums are being achieved. Back in 2006, I explained in the context of the UK market that amidst rising energy costs (and raw materials), the need for energy-efficient products in both new and existing properties is vital. I also was a proponent of consumer metrics and energy labelling to enhance consumers' decision-making on energy efficiency (at the time through EU EPCs). The property profession at the time still viewed sustainable buildings as niche and it was not always clear on the premiums that could be availed with the development of green buildings. Since then, things have changed and the pricing of property is a by-product of the upfront information we possess when making that buying decision. Therefore, sustainable assessment rating tools and labelling have become a big part of the drive towards more consumption of green buildings. With these global trends in mind let's turn our attention to the practicalities of measuring sustainability in Dubai and provide a range of case studies that highlight the developments that have taken place over the last ten years.

Sustainability metrics and assessments in Dubai

Metrics and the measurement of sustainability that inform decision-making, both financial and legislative, are key to embedding our industry, both consumers (buyers and occupiers) and producers (developers) to think greener.

Historically, Dubai has focused on the economic profit of project delivery. The Dubai Government are increasingly attempting to control the environmental standard by introducing a range of sustainable building codes, working alongside several international metrics. Box 11.1 highlights the sustainability metrics currently present in Dubai. The complexity and choice of indicators can be overwhelming, In order to make effective progress we need to standardize the use of sustainability indicators globally, in a similar way to which we have done in other areas of the profession (such as with international measurement standards (IPMS)). This is where I think Dubai has got the balance right. In previous research, I carried out in the UK, developers were overwhelmed by the sheer number of sustainability metrics that could be used to rate buildings and developments. Therefore, they could not attribute which metric was best to use or which metric a consumer would use to make a purchase decision. Suitable progress has been made in Dubai in establishing a common set of tools to measure sustainability (and importantly the core is internationally recognized). The use of indicators offers the opportunity to improve knowledge and decision-making in the local property market by providing tools for analysis and decision-making. The global challenges of climatic risk have also meant sustainability measurement tools are needed to measure the impact of property development itself. From a synopsis of academic literature, key criteria that must be considered when measuring sustainability include (Hollander, 2002; Plimmer et al., 2008).

Box 11.1 Sustainability metrics in Dubai

Dubai introduced the use of Al'Safat as a building classification tool for new development in October 2020, replacing the former Dubai Green Building Regulations and Specifications.

   The introduction of these revised codes is likely to have a similar impact as the legislative building codes we see in other markets, such as Part L in the UK (England and Wales). As a minimum requirement, all new building in Dubai need to meet the Silver Sa'fa, with building owners and developers able to achieve higher performance (Golden or Platinum Sa'fa).

   The two most common international metrics used in Dubai are LEED and WELL. A brief summary of these is:

- LEED: Buildings are classified on four tiers (LEED; Silver; Gold and Platinum), first developed by the US Green Building Council in 2000. The measure is widely adopted internationally, with 83,000 projects LEED certified across 150 countries.
- WELL: Buildings are classified on a range of wellness and health indicators (air, water, nourishment, light, movement, thermal comfort, sound, materials, mind and community). The 'WELL'

assessment focusses more on human aspects and their interrelationship with the building environment. Buildings are classified in tiers (silver, gold, platinum), first developed in 2014, with around 4,500 projects across over 60 countries (Savills, 2021).

1   Relevance and impact: the indicator must represent what is being measured or the target
2   Validity and availability: must be objective; statistically defensible; credible and reproduced affordably/consistently in the future
3   Simple: be understandable to the general public and policymakers
4   Transparent: the indicator must be clear and concise in its measurement; neutral and fair
5   Reliable and repeatable: to create the medium to long-term strategic goals required, based on a range of accessible data
6   Data should be easily and freely available
7   Cost-effective: data collection and analysis should not be prohibitively expensive
8   Comparability: general data sources or those widely accepted

If these criteria are met, then the indicator becomes a practical tool for the decision-maker and importantly provides a useful form of information for the user; a rational metric rather than one that creates distorted or biased behaviours. Most sustainability indicators in Dubai fulfil these ideals. However, what is important to add is the fact that sustainability is not now simply measuring energy or environmental performance, there are a range of more intangible, subjective measures that surround social and health and well-being, measures that are more challenging to consistently measure, for example, can we assess how productivity changes if we occupy green buildings?

Sustainability and its impact on the built environment are global problems. Dubai has made huge progress in moving the city towards a more sustainable future. The Dubai 2040 plan, as a notable reference, an abundance of green space and public realm infrastructure will only enrich the high-quality living standards Dubai already provides. The recent health crisis has placed emphasis on people and businesses to think about social enrichment and well-being. Addressed in the 2040 plan, Dubai will see the amount of green and recreational space double; 60% of Dubai's landmass will be allocated to preservation and nature, and social mobility is centre-stage with a range of new mobility routes set out. Dubai has also established like many other international markets, a green building standard (via building regulations, Al'Safat). Furthermore, Dubai was the first Arab city to be awarded a platinum rating for its city's certification (Gulf News, 2019).

According to the Emirates Green Building Council (2021), there are several specific targets centred on sustainable development, including the reduction of energy and water consumption in Dubai by 30% and the increase in the share of renewable energies to 25% – both by 2030. Dubai has also announced its Clean Energy Strategy to achieve 75% clean energy by 2050. With such clear guidelines in place, Dubai's government has increased its focus on sustainable buildings to comply with numerous global green building initiatives and regulations. Key initiatives in Dubai include:

- Dubai Plan 2040, addresses six focus themes, each highlighting a group of strategic developmental aims for Dubai: the People: "City of Happy, Creative & Empowered People"; the Society: "An Inclusive & Cohesive Society"; the Experience: "The Preferred Place to Live, Work & Visit"; the Place: "A Smart & Sustainable City"; the Economy: "A Pivotal Hub in the Global Economy" and The Government: "A Pioneering and Excellent Government".
- Dubai Municipality introduced the Al Sa'fat rating system in 2016 to strengthen the sustainable built environment in the city.
- "Trakhees" used for the issuance of building permits for construction and engineering works, including sustainability credits as part of its assessment.

The following case studies in Box 11.2 hope to shed light on the innovation in real estate development in Dubai that are exemplar to sustainability metrics.

Box 11.2 Dubai's most sustainable real estate

According to CORE (2017), Dubai has over 550 projects under LEED certification and over 75% of under-construction projects in Dubai are green-certified. According to the same report, government entities, and manufacturing and logistic occupiers are leading the charge in occupying green buildings. Below are five examples of a range of sustainable (high-rated) buildings in Dubai:

Case study 1 – Pacific Controls Headquarters Building

The project was opened in 2006 located in Techno Park, Dubai, and became the first sustainable building in Dubai (and was the first to receive a LEED platinum rating in the UAE). Key notable technologies include solar-thermal conditioning for the circulation of fresh air; water efficiency measures; solar panel sourced lighting and use of materials with a high recycling content.

Case study 2 – DEWA HQ

One of Dubai's first LEED Platinum buildings. This government office is home to Dubai Electricity and Water Authority (DEWA).

Case Study 3 – The Change Initiative (TCI), Sheikh Zayad Road

In 2013, TCI is a 4,000 sq m shop that provides sustainable solutions in Dubai. It has secured the LEED Platinum status and at the time was the most sustainable commercial building in the world (surpassing the previous top-spot The Pixel Building in Melbourne, Australia). A range of building technologies include: roof solar-panelling which provide up to 40% of the building's energy; water recycling; heat reflective windows; use of materials with a high recycling content and use of insulation (three times more than traditionally used).

Case study 4 – Dubai Sustainable City (Diamond Developers)

The first net zero residential community in Dubai was built in 2015. A 460,000 sq m residential development located on Al Qudra Road, an exemplar of a sustainable mixed-use community. Features include: 11 bio-domes (for urban farming); greywater recycling and solar panel energy sources, benefiting residents by having zero annual management service fees to pay. Mobility within the development is completely car-free with the use of community electric buggies. Within the development, there is a 50,000 sq ft institute for further research and development of sustainability practices (SEE Institute).

Case study 5 – ICD Brookfield Place (DIFC)

The Middle East's tallest and largest LEED Platinum certified office building, home to over 140,000 sq ft of green space as part of the development. Various reuse of materials throughout the construction phase (overall 50% of waste diverted from landfill), including re-use of existing piled foundations installed for a previous, abandoned development (Robert Bird Group, 2020). Set to be net-zero carbon emissions by 2030. According to JLL, ICD-Brookfield Place has increased fresh air by more than 30% above ASHRAE requirements and significantly above local standards. Uses recycled water to nourish the on-site green spaces and 40% reduction in water use through recovery technology. Advanced building management ensures efficient energy usage and provides shading and optimal natural lighting to boost the occupier's mindfulness and productivity The building commands a significant rental premium of at least 35% above the DIFC average (see Chapter 5)

Over time, property professionals recognize that sustainability is not just restricted to the presence of green ratings and energy labelling, it is more holistic with social, economic and now governance a key part of the future definitions. Two notable initiatives that stand out in Dubai include Expo City (District 2020) and Dubai Sustainable City. A brief overview of these is shown below:

Expo City Dubai

Type of Initiative:	A sustainable, tech-enabled, human-centric district of the future
Status:	Opening 1st October 2022
Objectives, Purpose & Aims:	Expo City is the legacy plan for Expo 2020 Dubai which ran from October 2021 to March 2022, and is set to be one of the key communities of Dubai that supports the realization of 2040 masterplan The main thematic focus as a new district include: sustainability; collaboration; knowledge-transfer; and talent creation.
Description:	Expo City Dubai is positioned as the blueprint for a technology-enabled (5G network) and sustainability-focused community. It will be Dubai's first WELL-certified community and home to 123 LEED certified buildings. Occupiers will benefit from smart meters to monitor and improve energy consumption and efficiency decision-making.
	Expo City Dubai also promises "meaningful educational and cultural experiences that inspire learning, agency and creativity, and instil a sense of purpose and fun across the themes of sustainability, mobility, and culture, carrying on the legacy of Expo 2020 Dubai". In delivering against its aims, there is a large focus on STEAM (Science, Technology, Engineering, Arts and Maths) which will be delivered through educational programs that are also focused on community engagement.
	In terms of business, Expo City Dubai will offer an evolutionary collaborative and tech-focused environment, in which large enterprises, government entities, SMEs, start-ups and academia co-exist and are able to utilize wellness amenities, hospitality and entertainment facilities.

Source: https://www.expocitydubai.com, supplied by Tim Shelton

The Sustainable City Dubai

Type of Initiative:	Integrated community in Dubai comprising 500 villas and 2,700 residents that are founded on sustainable and values-based living
Status:	Active community established in 2017
Objectives, Purpose & Aims:	The Sustainable City outlines its strategy as comprising a three-tiered approach based on the three pillars of sustainability (social; environmental and economic). A wide range of community outreach programs and operational savings through sustainable/smart technologies.
Description:	The Sustainable City offers an integrated and balanced approach to sustainable living and is hoped will act as a model for future community living. The award-winning community provides the following resources and facilities:

- Car-free residential living clusters with the provision of electric vehicles
- Environmentally friendly lifestyle with recycling and clean energy (10 MW/h of solar power is produced, where savings are passed back to residents).
- A focus on health and well-being with ample fitness and outdoor amenities
- 3,000 m^2 of urban farming space providing locally grown produce.
- 60% green spaces and two lakes of recycled greywater.
- Employment opportunities for residents.
- Social activities and community engagement to provide a "green education"
- The Sustainable Plaza – a collection of retail, office, business incubators and F&B outlets that are focused on sustainability, the circular economy and the sale of locally produced goods and services.
- Ability for residential investors to access "green residential loans" with preferential rates.

Source: https://www.thesustainablecity.ae/, supplied by Tim Shelton

The future: Can we create new sustainable business models for Dubai?

Given the global drive for ESG and sustainability, it is apparent that urban centres like Dubai play a critical role in developing and promoting more green practices. Dubai is benefitting from adopting metrics that are internationally recognized, reducing the friction on progress by having a multitude of different measurement tools. Larger developers like Emaar and Majid-Al Futtaim are making great strides in reporting their sustainability initiatives, which help educate and inform others to perform similar processes. Of course, sustainability is not merely about the existence of these metrics, it is about optimizing our current stock of property assets. In a recent panel discussion with Emirates Green Building Council, I presented some business models that highlight the institutional barriers that exist that prevent a more sustainable occupation and use of real estate globally. A summary of these discussions on occupational barriers include:

- Increasing flexibility within existing space: 50% of office space is inefficiently used, meaning that these buildings are not occupied to their full capacity. One solution to this issue would be to allow head tenants' rights to sublease and allow other businesses to use space within existing buildings. Building owners can be more flexible and buildings can be opened up for a more diverse range of uses so that there is less pressure to build more buildings.
- Creating adaptive and multi-functional assets/buildings: enhancing the cash flow, and adaptive space and be more reactive to changes in business demand and therefore providing a longer term, higher economic value within these adaptive new buildings.
- Data-driven decision-making: addressing the knowledge gap of "green buildings, grey users", that has meant tenants are not matching the environmental performance planned at the certification stage during the occupation state.

The Dubai Government is also making significant progress in its planning and legislative drive to make Dubai a leading green city. The literature contained in this chapter has only touched on the wide range of innovations that exist. In the future, Dubai has an opportunity to improve on what is being done and surpass many of its global peers. In 2021, Dubai went live with its sustainability building classification system (rated 1 to 4+), which has rated over 18,000 buildings in the Emirate, encouraging building owners to meet higher quality ratings related to sustainability features (Gulf News, 2021). A key feature of the newly introduced rating system is that reports can be viewed via the Dubai REST app, improving transparency and decision-making related to sustainability.

Local banks and financiers have also been involved in supporting sustainability, offering green home loans on properties that meet specific sustainability criteria. For example, HSBC began offering an interest rate discount of 0.25% and a 50% discount on applicable arrangement fees availed on LEED-certified properties (or similar). While not a significant saving, it indicates an intention for banks to start offering more favourable financial products to green buyers.

In relation to commercial properties, green leases or MoUs are evident, used as an institutional mechanism that closes the gap between building ratings/design and occupational savings (which historically has meant simply leasing a green building is not enough). A good example of this in practice is shown in retail developments of Majid Al Futtaim as well as occupiers to be at Expo City. Both examples focus on working with tenants to improve the operational efficiency of occupying tenants. This more prescriptive management through the lease closes the gap between performance of a building at the design stage and that at the operational stage, the latter of which is a much longer time duration, providing a much better chance of the buildings making an impact towards net-zero targets.

Another key area of sustainability within operational sustainability is the emphasis upon wellness and health and well-being. As previously discussed, the top costs for a business are its workforce and the workplace, therefore there are potentially greater savings to be made around occupying greener buildings, especially for more qualitative measures around employee productivity, talent retention and well-being. Sustainability is clearly emerging as a top priority for occupiers, but traditional factors such as location and accessibility remain more important (Livingstone and Ferm, 2017; Waters, 2006). Real estate developers need to find that balance and ensure ESG and traditional criteria for occupiers co-exist. New development like ICD Brookfield Place and Expo City present exciting opportunities to build from, showcasing it can be done in the Middle East.

Conclusions

In terms of a sustainable urban form, Dubai has been historically, quickly dismissed, largely by a high level of car dependency and limited forms of urban mobility. Yet, these public transit systems have continued to expand, initially with the operation of Dubai Metro since 2009 and the subsequent introduction of a tram system within the inner, more dense parts of Dubai. Global planners today observe mixed land use models as imperative to urban sustainability. The presence of high-density residential development with supporting retail, business premises and services is a key part of the solution, and Dubai has a plentiful amount of these. Diverse housing tenure is also a major attribute. Mixed-use buildings are also concepts that have been developed and worked successfully in Dubai, where offices, retail, residential and hotel sit within a singular building.

As I have discussed in earlier chapters, occupiers are demanding more flexibility via short leases, spurring a more active feedback loop for property owners in terms of measuring obsolescence and sustainability. Building owners will be driven to perform to higher metrics and rating systems and labelling becomes an enabler for change and improvements in environmental and social outputs.

As the market is increasingly supplied with more and more information on sustainability and performance, valuers and strategic consultants will be expected to be the "cheerleaders of change", communicating the true impact of sustainability KPIs to building owners and investors. They will play an ever-important role to showcase that ESG is something that can no longer be a secondary consideration. Dubai demonstrates successful implementation when it comes to providing sustainable buildings. Undoubtedly, its future urban form will be home to more. Dubai 2040 is promoting sustainability to be at the forefront of future development, focusing on equal measures in the social provision of parks, green spaces and other public realms.

The final chapter of this book will examine somewhat of an interrelated theme, smart cities, enabling a more sustainable urban form through enabling technology.

References

Cadman, D. (2000) 'The vicious circle of blame', cited in Keeping, M (2000) *What about demand?Do investors want "sustainable buildings"?* London: Royal Institution of Chartered Surveyors..

CORE (2017) The Green Issue: Sustainability and Wellness in Dubai. https://www.businessimmo.com/eu/research/85701/the-green-issue-sustainability-and-wellness-in-dubai-2017.

Dixon, T., Pocock, Y., and Waters, M. (2006), 'An analysis of the UK development industry's role in brownfield regeneration', *Journal Property of Investment and Finance*, 24 (6), pp. 521–541.

Emirates Green Building Council (2021) 'UAE Sustainability Initiatives' https://emiratesgbc.org/uae-sustainability-initiatives/. Accessed June 2022

GulfNews(2019)DubaiReceivesPlatinumRatinginLEEDforCities.https://gulfnews.com/business/dubai-receives-platinum-rating-in-leed-for-cities-1.1555540236990.

Gulf News (2021) Dubai Introduces 1–4 Star Ratings for Its Buildings - and There Is a '4Plus' for Sustainability Too. https://gulfnewscom.cdn.ampproject.org/c/s/gulfnews.com/amp/business/property/dubai-introduces-1-4-star-ratings-for-its-buildings---and-there-is-a-4plus-for-sustainability-too-1.1636946954451.

Hollander, J. (2002) Measuring community: Using sustainability indicators'. In Devens MA Planners Casebook, 39, Winter 1–7.

Innovate UK (2016) Building Performance Evaluation Programme: Findings from non-domestic projects, Getting the best from buildings. https://www.ukri.org/wp-content/uploads/2021/12/IUK-061221-NonDomesticBuildingPerformanceFullReport2016.pdf

Leskinen, N., Vimpari, J., and Junnila, S. (2020) 'A review of the impact of green building certification on the cash flows and values of commercial properties', *Sustainability*, 12, pp. 1–22.

Livingstone, N., and Ferm, J. (2017) Occupier responses to sustainable real estate: what's next?, *Journal of Corporate Real Estate*, 19 (1), pp. 5–16.

Majid Al Futtaim. https://www.majidalfuttaim.com/en/media-centre/trends-and-insights.

Plimmer, F., Pottinger, G., Harris, S., Waters, M., and Pocock, Y. (2008) *Knock it Down or Do it Up?: Sustainable House Building: New Build and Refurbishment in the Sustainable Communities Plan*, Watford: BRE Press.

Lorenz, D. P. (2008) 'Breaking the Vicious Circle of Blame – Making the Business Case for Sustainable Buildings', *FIBRE*, Brussels: RICS Research

Robert Bird Group (2020) ICD Brookfield Place: Dubai. https://www.robertbird.com/rbg-projects/icd-brookfield-place/.

Savills (2021) ESG in the Middle East, Savills Research 2021. https://research.euro.savills.co.uk/uae/dubai/final---esg-me-report.pdf

Waters, M. (2006) 'Environmental sustainability: Is it HIP?' *RICS Land Journal* (November 2006) p. 4.

Waters, M. (2007) 'The great energy debate', *RICS Land Journal* (January 2007), p. 8–9.

World Economic Forum (2022) The Global Risks Report 2022. https://www.weforum.org/reports/global-risks-report-2022.

UNEP (2021) *United Nations Environment Programme: Global Status Report for Buildings and Construction: Towards a zero-emission, Efficient and Resilient Buildings and Construction Sector.* Nairobi

WCED (World Commission on Environment and Development) (1987) Our Common Future: Report of the World Commission on the Environment and Development [Bruntlandt Report]. General Assembly, published as Our Common Future. 1987. Oxford and New York: Oxford University Press.

12 The future of Dubai as a Smart City

Chapter 1 reviewed the economic and urban development of Dubai over the last 50 years. Be it agile and visionary leadership, the introduction of the city-as-a-brand concept, mobilization of large pools of global talent, the desire to transition away from a hydrocarbon-based economy, the ability to harness advances in technology and innovation made possible by public-private sector collaboration, or well-defined governance and policy informed through stakeholder consultation; Dubai has risen from the desert and got to where it has, fast – predominately due to the effectiveness of the enabling environment it has put in place to fuel such phenomenal growth.

But what is in store for Dubai over the next 50 years? If the last half a century is anything to go by, then be prepared! We must consider though, that the world is a very different place from when its journey began. Dubai will therefore evolve in a manner shaped by a very different set of global trends and drivers of growth in an ever-connected and virtual world. Whilst divisions, attitudes and disparities still proliferate across the planet, the one thing that every global citizen and nation in the world have in common is the exposure to the many complex challenges. The previous chapter highlighted the climatic risks. On a more responsive note, the world is also now more digital. Technological advances have demonstrated that geographical barriers no longer present the challenge they once did, especially pertinent in terms of access to global skills and talent. Furthermore, the capabilities of virtual working, highlighted by the onset of COVID-19, have also been proven, meaning that hybrid working practices and flexible workplaces are becoming normalized. Such trends will undoubtedly continue to have a profound impact on the way in which cities and urban landscape form. Cities have the potential to be centres of innovation, attracting global talent that will help solve the world's greatest challenges – the success of them doing so being a key differentiator in terms of global city competition and a city's ability to attract foreign direct investment and skills. On the ground, there will in parallel also likely be an ever-evolving "blurring" between physical and virtual spaces, powered by immersive technology and the evolution of new technological phenomena such as the metaverse. Predicting the future of Dubai is therefore an exciting, but extremely complex exercise. Much is

DOI: 10.1201/9781003186908-16

said and written about the rise of the so-called "Smart City", but what does this label mean, and can it be used as a lens through which we can view and chart Dubai's next 50 years?

This chapter will begin by contextualizing the term "Smart City" to enable a better understanding of the opportunities and challenges that Dubai will face over the next half-century. It will assess various components of its completed, planned and inflight growth strategy against selected key drivers of success. The chapter will culminate in the proposal of a number of opportunities that Dubai has to differentiate and cement itself as a top-tier global city, including specific reference to opportunities within real estate.

This chapter will be structured as follows:

Section 1	**Contextualizing Smart Cities**: A definition, the challenges, a means of differentiation, performance assessment and archetyping
Section 2	**The Happy City concept** – The role and importance of human-centricity and happiness as a key pillar of a Smart City vision
Section 3	**Smart City drivers and enablers** – Going beyond output-focused Smart City assessment tools – What are the key drivers and enablers in the evolution of the Smart City and the happiness of its residents?
Section 4	**Dubai's future-focused strategy** – How is Dubai fairing against the drivers and enablers put forward in Section 3 in terms of delivered and in-flight initiatives and what are the challenges it faces in delivering its vision?
Section 5	**Opportunities for differentiation** – What could be the key areas of differentiation for Dubai on the world stage as it evolves over the next 50 years?

Contextualizing smart cities

The term Smart City has become somewhat of a buzzword over the last 5–10 years it seems, with many different interpretations and opinions being offered both in and out of academia. With access to innovative technological resources and seemingly no limit on what can be achieved through human endeavour, the term Smart City has become a catch-all term or umbrella concept that embodies ambition, connection, optimized decision making and means of differentiation on the global stage that results in outcomes that improve quality of life, termed as *"a concept viewed as a vision, manifesto or promise aiming to constitute the twenty-first century's sustainable and ideal city form"* (Trindade et al., 2017).

In an ever-connected world, the rise of the Smart City can also be viewed as intrinsically linked with the dawn of "Industry 4.0", an era defined as the *"realization of the digital transformation of the field, delivering real-time decision making, enhanced productivity, flexibility and agility"* (IBM, 2022). This is somewhat of a technologically-centric view but underlines the importance of advances in connectivity led by technologies such as automation, cloud computing, big data, autonomous robots and artificial intelligence (AI) – the so-called "smart" technologies that are disrupting industries, changing the face of how humans work, live and play and, therefore, having a profound impact on urban, physical forms and city evolution as a result. Most recently, Smart Cities have been called out in terms of their response to the recent COVID-19 pandemic and their ability to respond swiftly and effectively through an innovative use of technology. Dubai has been heralded for its efforts in responding to the pandemic through data-driven responses focused on highly accessible COVID testing, vaccination program, people-movement monitoring and the unification and highly effective coordination of healthcare stakeholders. This resulted in Dubai climbing in the IMD Smart Cities Index rating from 42nd in 2020 to 28th in 2021 (IMD, 2021). The principle of resiliency is therefore also a key characteristic of Smart Cities, the ability to respond to "shocks", and in this instance through tailored human-centred responses, powered by technology and data-driven decision-making. The pandemic has seen an acceleration of digital and ecological transformations in smart cities and *"this acceleration is redefining resilience, which is increasingly becoming a local objective. And cities that have been seen as handling Covid-19 challenges in an efficient and effective way rank high in the report"* (IMD, 2021).

The size of the Smart City market and the impact of market dynamics

According to market reports, the current value of the global Smart City market stands at around USD1 trillion and is forecast to grow at around 25% CAGR per year leading to a market value of around USD7 trillion by 2030 (Grandview Research, 2022). But what are the predicted returns for this significant investment? In a study undertaken by the McKinsey Global Institute, it is claimed that notwithstanding the benefits driven through non-technological initiatives, the following example benefits maybe realized purely through the effective utilization and implementation of city-wide digital applications alone (McKinsey Global Institute, 2018):

- 30–300 lives saved each year in a city of 5 million;
- 30%–40% fewer crime incidents;
- 8%–15% lower disease burden;
- 15–30 minutes shaved off the daily commute;
- 25–80 litres of water saved per person;
- 20%–35% faster emergency response time

With the size of the market being such an attractive opportunity for profit growth in the private sector and with the estimated benefits being perceived as being so tempting by city governments, the risks associated with the marginalization of the ordinary city citizen are heightened, especially given the perceived gaps in capability in contemporary skills and knowledge at government level, that city leaders perceive as the drivers of differentiation (i.e., blockchain, cybersecurity, AI, machine learning, etc.) culminating in a reliance on private enterprise. Indeed, in a piece written for the World Economic Forum (WEF) in 2019, it was proposed that the stakeholders of Smart City projects are politicians, consultants, academics and tech companies. However, the most important group of stakeholders is often missing: the ordinary citizens that will have to live in these transformed cities (Weber, 2019). But what does all this mean and how can cities, such as Dubai, adopt a balanced, sustainable approach to city development and evolution that involves and responds to the needs of everyday city citizens and balances the objectives of the public and private sectors?

However you choose to view or interpret the concept of the Smart City, at a national level it could be argued that they can be observed as a means of strategic competitive advantage and a conduit for attracting human capital and increasingly mobile global investment – an evident strategy that the city of Dubai has deployed in such an innovative and effective manner; more of which will be demonstrated and discussed later in this chapter. Smart Cities also have to be viewed through the lens of the challenges and trends that they seek to overcome, if we are to truly understand the drivers of their evolution in the future and what city outcomes could look like. These challenges exist at city/national, regional and global levels.

The challenges

According to numerous reports and research, it is estimated that in 2050, approximately 70% of the global population will reside in cities. This figure compares to around 50% currently (UNEP, 2018). This represents an increase of approximately 2.5 billion people or approximately 1.5 million people per day that will be added to the global urban population requiring urban areas to expand and new cities to emerge. In addition to the challenges driven by urbanization at city level, planners, policymakers, government and city stakeholders will also have to contend with an array of global trends that will influence and shape cities over the next 50 years. Some of these key global trends can be categorized as follows:

1   Cities rising and redesigning
    • Growing demand for seamless multimodal transport
    • The growing primacy of data-driven predictive city planning
2   Demographic change and migration
    • Data transparency and accessibility
    • Ageing populations that place greater pressure on healthcare systems and place more of an acute focus on the provision of facilities for the elderly

3 Climate change and energy transition
  - A changing climate requires infrastructure adaptation
  - An increased focus on renewable energy and preventing CO_2 emissions
4 Digitization, cyber security and AI
  - Blurred lines between physical and digital space and the growing prominence of the metaverse
  - Data and trust
  - The impact of digital technology on our lives – the importance of "disconnection" and community engagement
  - Disrupted retail disrupts the city
  - The emergence of AI
5 Natural world in decline
  - Strengthening our relationship with nature
  - Declining resources on our natural world
6 Globalization and geopolitical shifts
  - Global citizenship
  - City governance on the brink of disruption
7 Individual empowerment
  - The rise of the e-citizen
  - (in)equality in the city
  - The co-dependence of universities and cities
  - The war for global talent
8 Well-being 2.0
  - Healthcare and well-being
  - A shift to local food production
  - Loneliness is the new illness
  - New modes of living
  - Universal design to accommodate all in the city

Given the significant and varied challenges and trends, therefore, it is not surprising that there is a generally held consensus that cities need to drive the change needed to realize a secure, resilient and sustainable future for global citizens. Indeed, the United Nations have specifically called out the need for sustainable cities and communities as part of their 17 Sustainable Development Goals, and more specifically underlined the urgent requirement to, make cities and human settlements; inclusive, safe, resilient and sustainable (United Nations, 2022). The urgency is heightened as the UN further state that whilst 156 countries have developed national urban policies, only half are in the implementation stage, meaning there is still a lot of work to be done.

Cross-regional collaboration: Dubai leveraging its position

There is, therefore, a significant opportunity for Smart Cities to drive benefits beyond their own urban boundaries if collaborative, unified approaches can be adopted at both a macro (regional) and global level. Smart Cities, in many ways, have an opportunity to act as nodes where innovative thinking and solutions could be scaled to benefit a greater number of people and society at large. For example, and more specifically, there has been a growing argument for greater attention and investment to be made to enable enhanced integration across the Middle East, North Africa and Pakistan (so-called "MENAP" region in which Dubai is positioned), to enable the leverage of scale economies to enhance GDP for constituent nations. In a report released by Majid Al Futtaim (a Dubai-based, major retail and lifestyle brand), McKinsey and the WEF, it is stated that, while 8.5% of the world's population lives in the MENAP region, it only accounts for 3.4% of global GDP. An incremental USD2.5 trillion would need to be generated for the region to produce its fair share of GDP. Ultimately, however, the region continues to be more fragmented than others, with an underdeveloped private sector compared to global benchmarks (MAF, 2022).

Whilst the underlying political, social and economic obstacles and challenges that constrain cross-MENAP integration are beyond the scope of this chapter, the very real opportunity for more successful cities within the region to leverage their position and "smart" status to support such transformation is worthy of mention. Indeed, the MAF report further outlines that Dubai's strength in attracting foreign direct investment (USD7 billion in 2021, a 5.5% increase from the previous year, placing the city as the most successful globally), represents a significant opportunity to unlock the benefits associated with greater collaboration and coordination that will further evolve the private sector.

The future development of sustainable, agile and resilient cities is therefore characterized by significant complexity. It is clear that a great deal of collaboration between an array of different stakeholders will be required. Indeed, the role of the private sector and corporate enterprise working alongside an agile city government should not be underestimated in terms of the planning, design, financing, roll-out and scaling of city-wide solutions that aim to enhance the quality of life for their citizens. However, some claim that a degree of caution should be observed when there is such a reliance on the private sector which may be motivated to employ opportunistic behaviour to drive profit, especially when the estimated size of the Smart City market is perceived to be so significant.

Smart City development pathways and the importance of people-centricity

One principle that is growing in prominence, is a focus on the human, or so-called people-centric approach to city development. As Zakzak (2019) suggests, whilst developing a Smart City can follow numerous pathways (i.e., technology, data, environment), following a people-centric Smart City developmental path requires a societal approach that involves public engagement and participation as a core developmental philosophy, that requires the city to develop the capacity to widely measure well-being and the state of "happiness", and respond through its public policy frameworks. Pertinently, people-centricity or city happiness is the cornerstone of Dubai's Smart City vision. Indeed, the city has the ambition to become the world's happiest city as it strives to differentiate itself on the world stage and be a key driver of the UAE's 2071 Centennial Plan. The concept of happiness is explored in more detail in the next section.

Dubai – The world's happiest city

Policymakers obviously see that the key to a successful, thriving and resilient Dubai is putting the happiness of its residents and visitors as the core pillar of their city evolution strategy. Indeed, although acknowledging technology and innovation as key levers that can be pulled, there is an emphasis on putting people's wants and needs at the forefront. Effectively Dubai policymakers believe that this will provide valuable insight that will direct and provide the "burning platform" for city initiatives, powered through technology and private/public sector collaboration, that will ultimately improve people's lives. Dubai's strategy is to raise and sustain happiness levels across all aspects of residents' and visitors' experience of the city – a strategy, the city believes will genuinely differentiate them on the global stage. The acknowledgement of happiness as a key part of City strategy is not specific to Dubai however, indeed, more generally it is also observed as being a contemporary indicator of social progress by governments and global organizations including the United Nations. The first "World Happiness Report" was published in 2012 and as a consequence, several countries, including the UAE have led the way on investing, developing and nurturing social well-being. But can such lofty ambitions aimed at becoming the world's happiest city be viewed merely as an attempt to market and brand the city, without any real foundation? After all, is happiness not subjective? Dubai disagrees. They cite their happiness agenda as a one-of-a-kind, scientific and systematic approach to improving the lives of their residents and visitors by putting people at the centre.

Happiness drivers

In a study conducted by Zhu et al. (2022) and in response to what they state is an insufficient attention paid to investigations on how the Smart City acts on human happiness, they propose a structured and systematic view in regards to the impact of adopting a human-centred approach to Smart City Development. Their framework is based on four key pillars of the Happiness-Driven Smart City ("HDSC"), that include: efficient and green infrastructure; innovative economy; inclusive society; and sustainable environments with each pillar being contributed to by a number of key Smart City themes or focus areas. Clearly, their view on Happiness and the Smart City is one with a broad perspective that embodies not only traditional, tangible measures of progress such as physical spaces and infrastructure but also wider cultural themes such as a spirit of innovation and inclusivity that are more difficult to measure.

Adopting a more holistic view is important given the current rate of change, the nature of the challenges and trends we face now and in the near future (such as those highlighted in the previous section) and the ever-evolving needs of the city citizen. It is clear that if Smart Cities are to evolve in a manner that is sustainable and genuinely puts the needs of its citizens at the forefront, they cannot rely on technology alone. Attitudes, physiological needs, equality and inclusivity are becoming increasingly important and becoming recognized as key contributors to the enhancing social value.

Drivers of city residents' satisfaction and happiness

In addition to academic literature, recent commentary from a leading strategic advisor Boston Consulting Group (BCG) also supports a view in regards a holistic perspective on the impact of Smart City Happiness. Their own analysis demonstrates six reasons that make residents' satisfaction and happiness increasingly important for cities. These six reasons are highlighted in Box 12.1.

Box 12.1 Key reasons why happiness is important for cities

1 Attracting & Retaining Talent – The perceived levels of liveability and overall happiness of a city have a direct impact on the ability it has to attract mobile talent to (re)locate to it and keep it there. Ultimately, talent is the most important driver of innovation and is the "engine-room" of city evolution.
2 Resource Health – Happiness is connected to a healthy lifestyle and longer life, which BCG outlines, reduces pressure on the healthcare system and keeps talent performing at their optimal level.
3 Project Support – Cities need their inhabitants to support major capital investment programs that act at the catalyst for evolution. This can only be achieved through the articulation of a compelling and forward-looking investment case with which inhabitants can resonate, identify the benefits and be compelled to support.
4 Conflict Avoidance – In a rapidly changing world, differences may occur between social groups with regards their opinions of the use

and development of urban spaces. City governments need to be aware of the potential for such conflict and put in place measures to mitigate such risks.

5   City Democracy – It is cited that happy city residents are more likely to participate in city government elections.

6   Evolving Happiness – It is proposed that, "happiness creates happiness" and that the proliferation of caring social relationships within the city can support a virtuous cycle of city-wide happiness.

Source – Aguiar et al. (2020).

Measuring happiness – Dubai's government's interaction feedback

One key initiative that the government of Dubai is placing a significant focus, and which highlights the importance of citizen, government engagement by way of emulating the spirit of community and to feel empowered to make a difference and to be part of a city's evolution, has been their relentless focus on capturing customer feedback on interactions with government and quasi-government entities. Whether it's interacting with the Dubai Electricity and Water Authority ("DEWA") having just paid a bill online, using your online banking application to undertake a money transfer, or having just touched down in Dubai having taken a flight from London on Emirates airline; requests for feedback via a 1–5 rating will quickly follow. The resulting value of the data is seemingly in its scale. Through the combination of datasets from multiple client profiles and data from high-frequency touch points, the insights harvested from this example of societal big-data collection can then be used to inform resulting policy, and/or innovative responses that provide the consumer with the confidence that their needs are being addressed and trust nurtured. Indeed, as Zakzak (2019) outlines, there is a specific need for a city to develop the capacity and to widely measure well-being and the state of "happiness", and respond through its public policy frameworks.

Moving towards a balanced set of smart city drivers and enablers

So far it has been identified that the challenges, trends and drivers that have traditionally shaped Smart City Development over the last 50 years have changed and as a result, will likely lead to urban development strategy adopting a very different form over the next 50 years. In a new era, defined by connectivity, the benefits of a successful Smart City strategy are not limited to that city, there are significant opportunities for knowledge transfer and scaled spill-over effects to reach regions and the world in turn. Effectively, a greater number of people will benefit if cities exploit new network effects – a key principle if we are to overcome the most difficult global challenges facing us today.

It has been identified that adopting a balanced view towards the analysis of what constitutes success and performance of Smart Cities is vitally important. Indeed, there is no one-size-fits-all approach to be used to position and benchmark what success looks like, despite the numerous publicly accessible Smart City indices that seem to want to do so.

Simply put, the evolution of Smart Cities will not solely be dependent on the delivery against technologically and data-driven use cases, instead, success will likely be predicated on the extent of effective governance and resulting responsive policy that effectively engages the city citizen and continually gathers feedback from highly representative, and inclusive information gathering programs, and also continues to engage the citizen in the development of responses. Indeed, as Senge (2006) outlines, governance plays an undoubtedly substantial role in making and maintaining the sustainability of smart cities. A Smart City is, in an organizational sense, a learning organization, and governance needs to adopt the learning organization principles (Senge, 2006).

In short, one of the key sets of principles that will be pivotal in driving successful Smart City strategy will be that city's ability to engage, learn, test, validate and respond.

The next section will position a balanced set of drivers and enablers that will likely shape Smart Cities in the future.

Smart City drivers and enablers

As the concept of the Smart City has grown and gained greater prominence in the study of city evolution over the last 5–10 years, so too it seems to have a number of indices that aim to measure the success of Smart Cities, and as a result drive competition between them. These indices, offered by a variety of individual and sometimes hybrid mix of stakeholders including private sector companies, academic institutions, NGOs and professional bodies; vary in terms of their rankings and measures, seemingly as a result of a lack of consensus around an accepted definition of what Smart Cities are and their purpose. Indeed, in a study by Patrão et al. (2020) focused on the proliferation of Sustainable City Assessment tools ("SCAs"), they outline the following deficiencies:

1   There is a lack of balance distribution of indicators;
2   The great majority of the SCA tools are static assessments or snapshots;
3   They present limitations when comparing cities with different scales;
4   There is a lack of focus on the measurement or evaluation of the impact of city development and initiative goals and benefits;
5   The extent of stakeholder engagement in the articulation and delivery of city developments and initiatives is not assessed;
6   The contribution and/or alignment to the United Nations Sustainable Development Goals ("SDG's") is not assessed; and
7   The feasibility and robustness of the investment case(s) of city implementation is not acknowledged.

Building on item 1 of their list of deficiencies above; many of the published SCA tools and indices that exist today appear excessively technology-focused and do not consider human-centricity and effective city governance as success factors – key principles that have been discussed previously in this chapter.

The development of a single tool, or "one-size-fits-all" set of assessment criteria able to overcome all the mentioned gaps, could be considered to be quite an ambitious task, especially given the complexity of the challenges cities are faced with and the changing landscape of resources that are accessible to help overcome them.

As an alternative, and building on the principles presented in so far this chapter, a set of five Smart City enabling themes, each with its set of sub-factors, are proposed below. These criteria will subsequently be used as a framework to assess the strategies and initiatives that Dubai has implemented to deliver against its Smart City vision.

Drivers and enablers of smart cities

1 **Governance, regulation and incentivization**
    a **Government vision and leadership** – Is a governed, publicly communicated Smart City vision in place that contains clear and measurable objectives, metrics for success and defined milestones? Is the vision sponsored by the highest level of government and structured in a manner in which it can be cascaded through multiple value chains?
    b **Agile and innovative government** – The responsiveness of city-wide governance and the agility and speed of decision-making. To what extent is innovation institutionalized within government entities and supported through an integrated system of modern tools? Is legislation allowed to evolve quickly through adopting an agile approach to regulation development which, in turn, enables the government to grant licenses and therefore enable the rapid adoption of new innovative solutions?
    c **Incentivization of private sector innovation** – These are measures in place to encourage private sector innovation through the establishment of innovation and scientific research centres, adoption of new technologies and development of innovative products and services.
    d **Innovation through start-up incubation** – Establish a stimulating environment for innovation in the form of supportive institutions and laws and provide opportunities for cross-pollination of ideas through start-ups, corporate enterprise and government.
    e **Financing** – These are incentives in place to allow for private sector financing in planned Smart City initiatives and the innovations that support them.

2 **People and skills**
   a. **Talent retention incentives** – The presence of incentives (not solely financial) for attracting and retaining specific skills that are deemed to contribute to the strategic vision of the city.
   b **Purpose** – The next generation of leaders will be motivated by purpose – to feel that they are contributing to something. Does the city have a vision and strategy that is purpose driven and acknowledges the need to resolve global challenges and improve quality of life?
   c **Access to digital tools and resources** – Does the city have measures in place that target equal and fair access to digital tools and resources for its citizens?
   d **Balanced education that nurtures innovative thinking** – Does educational curricula (from early years through to further education) place a focus on the incubation of highly innovative skills by concentrating on science, technology, engineering, art and mathematics? To what extent are soft skills such as emotional intelligence (EQ), critical thinking, diversity and cultural intelligence and the ability to embrace change, integrated into curricula requirements?
   e **Educational partnerships** – Are there mechanisms in place to enable partnerships between the public, private and educational sectors (an integrated educational value chain) to enable knowledge transfer and innovation incubation?

3 **Civic engagement, resident-centricity, partnerships and trust**
   a **Resident-centricity** – Do city strategies and initiatives genuinely integrate the notion of resident-centricity? Are the expectations and needs of people at the fore in order to guide strategy and master planning, community management and the design of services?
   b **Transparency and ease of government interactions** – To what extent are interactions with the government transparent and of relative ease? Are government interactions routinely assessed for effectiveness and are citizens encouraged to provide feedback?
   c **Public/private sector collaboration** – Is an enabling and government-sponsored and coordinated environment in place that allows for co-creation and collaboration between public, private, institutional and educational stakeholders? Is there a focus on the cultivation of appropriate business cultures and shared objectives across this broad set of stakeholders?
   d **Citizen participation** – Are citizens actively encouraged to participate in Smart City programs and agile solution development? Is there a balanced, inclusive means of citizen consultation in place that considers varying levels of digital literacy, including online and offline data collection?
   e **Regional and global collaboration** – Does the city display a desire to share knowledge and participate in initiatives across regions and at a global level in order to leverage scaled capabilities?

4   **Technology and data**

    a   **Open data platforms** – Does the city provide accessible and connected urban data platforms that are supportive of the creation of new city-focused value propositions and data-led innovation?

    b   **Data governance and security** – What is the extent of protection of city-wide data infrastructure, is there a suitable level of investment being made to reduce the impact of potential cyber-attacks, data leakages and misappropriation of data?

    c   **Focused and scaled technology adoption** – Do city governments have specific strategies and initiatives in place to incentivize, support and scale specific emerging cross-sector technologies such as blockchain, 3D printing, and AI,?

    d   **Technology adoption measures** – Is technology adoption routinely measured and outcomes (in terms of both business and social value) assessed?

    e   **Agile approach to new technology regulation** – Is regulation and guidance in place to support the public, private and educational sectors in the pragmatic, responsible and safe implementation of emerging technology, i.e., AI?

5   **Focused problem solving**

    a   **Legacy capability leverage** – Are capabilities and investments that have historically shaped the city being leveraged and evolved for strategic and competitive advantage?

    b   **UN sustainability goals** – Is the future city vision and resulting initiatives supportive of the UN Sustainable Development Goals?

    c   **Community focus** – Is there a focus on the specific needs of local city communities and on cultural preservation? Are innovative and sustainable solutions and capabilities developed at the community level given airtime and supported by the government for scale?

    d   **Structured innovation** – Are measures in place to ensure that Smart City initiatives are properly assessed for their desirability, feasibility, viability and adaptability from vision through to execution?

    e   **Open innovation** – Are platforms and initiatives in place that enable joined-up thinking/collaboration and public participation?

The next section will use the framework outlined above as a lens through which a number of selected key strategies and initiatives, implemented in Dubai, can be assessed – the outcome of which will be used to build a picture of the potential future Smart City landscape and opportunities it has to differentiate itself on the world stage.

Dubai's future-focused strategy

It will come as no surprise that in a city that has vision and ambition seemingly indoctrinated in its DNA, there is no shortage of initiatives and strategies that aim to build upon the impressive progress of the last 50 years and position Dubai as the world's leading and most future-focused Smart City. But given the current and predicted global trends, challenges and opportunities that the world faces, how achievable is this vision and does Dubai have what it takes to evolve itself in a manner that is sustainable, relevant and capable of attracting and retaining increasingly mobile talent?

This section identifies and summarizes a selection of inflight strategies and initiatives (sourced from publicly accessible, and predominantly online resources) at governmental (national/city), sectoral, educational and community levels that have been designed and implemented to support Dubai's vision to be one of the world's leading Smart Cities and happiest city on earth. These will be viewed through the lens of the framework outlined in the previous section in order to decipher some of the challenges and opportunities that Dubai will need to address if it is to successfully differentiate itself on the world stage.

As one can imagine, in a city that never seemingly stands still and one with such lofty ambitions, it is impossible to acknowledge every initiative that will support the evolution of the city over the next 50 years. Instead, the initiatives presented have been deliberately chosen to provide the breadth of case studies necessary to most effectively enable an analysis using the framework proposed.

Figure 12.1 summarizes each initiative with a high-level assessment of outcomes given the maturity of each.

A synopsis of each tier of Smart City initiatives will be discussed below.

National-, governmental- and city-level initiatives

As can be seen from the selected case studies above, Dubai has no shortage of inspiring, policy-driven, top-down, initiatives that will guide and enable the evolution of Dubai in a strategic and increasingly inclusive fashion. From a guiding long term national vision to an urban masterplan, digital inclusion policy, special purpose vehicles for accelerating digital transformation and initiatives that encourage better access for financing, Dubai has evidently invested time and effort in ensuring that it has a balanced approach to governance and regulatory policy, whilst retaining its focus on remaining agile to be able to effectively respond to global challenges and leverage future opportunities.

The key challenges impacting the effective implementation of such initiatives and strategies, however, are predominately associated with their infancy and abundance. There will undoubtedly be a need to continually refine and revisit cited strategies and initiatives given likely changes in the external environment in which they operate. Given finite government resources, this will require a specific focus on private sector collaboration and strategy alignment, genuinely being able to engage citizens and the ability to

Dubai Evolution Factors	National, City & Gov't Level									Sector Level					Education & Cultural Level			Community Level	
	UAE Centennial 2071	Dubai 2040 Masterplan	Gov't with You	MBR Majlis	Digital Incubation	Future Foundation	Digital Dubai	Invest in Dubai	DIFC Finance Lab	AI Waste Energy	MBR Solar Park	RTA - Innovation	AviationX Lab	Centre for Smart Construction	RTA - Innovation Centres - Uni Birmingham	National Tolerance Program	The Sustainable City	Expo District 2020
Governance, Regulation & Incentivization																		
Government Vision & Leadership	o	o	o	o	o	•	o	•							•			
Agile & Innovative Government	o				•	o												
Incentivization of Private Sector Innovation					•	o				o								o
Innovation through Start-up Incubation					•	o		o	o		o	o	o					o
Financing				o	o		o	o	o	o	o	o						o
People & Skills																		
Talent Retention Incentives					•	o								o	o			o
Purpose	•	•	•	•	•	•	•	•	•	•	•	•	•	•	•	•	•	•
Access to Digital Tools & Resources				o	•	o								o	o			o
Balanced Education/Innovative Thinking											•			o	o		•	o
Educational Partnerships				•						•				•	•		•	o
Civic Engagement, Resident-centricity, Partnerships & Trust																		
Resident-centricity	o	o	o														•	o
Transparency & Ease of Govt Interactions	o	o	o	o		o												
Public / Private Sector Collaboration				•								•						
Citizen Participation	o	o	o	o							o	o	•				•	o
Regional & Global Collaboration				o	o		o					•						
Technology & Data																		
Open Data Platforms				o														
Data Governance & Security				o														
Focused & Scaled Technology Adoption				o	o						o			o	o			
Technology Adoption Measures				o							o							
Agile Approach to New Technology Regulation				o														
Focused Problem Solving																		
Legacy Capability Leverage									o	•	•	•			•	o	•	o
UN Sustainability Goals	o			o					o	•					•	o	•	•
Community Focus	o		o	o										o	o		•	o
Structured Innovation			•	o							o			o	o			
Open Innovation	o		o								o			o	o			o

Established initiative – Evidence of tangible outcomes against enabling factor	•
Semi established initiative – Evidence of partial outcomes against enabling factor	o
Initiavite in its infancy – tangible outcomes expected	o

Figure 12.1 City initiative outcomes.
Source: Supplied by Tim Shelton.

scale capability cross-sector, cross-region and vertically within industries. Suggested challenges and opportunities owned at a governmental and city level are summarized as follows with responses proposed in the final section of this chapter.

- **Encouraging greater citizen participation** – Genuinely inspiring city-citizens to participate in open innovation and policy co-creation through the utilization of platforms such as MBR Majlis and *My Government with You,* and placing more emphasis on the need for public consultation and the incentives for doing so.
- **Cross-regional market co-operation** – Showcasing the opportunities to scale and monetize solutions aimed at improving quality of life in a region that is currently disintegrated and placing barriers to investment, rapid growth and scale compared with other regional, continental markets such as North America and the EU.
- **Accelerating sectoral transformation** – Suitable integration between government and quasi-governmental/corporate entities for a specific sector-focused transformation (i.e., construction and real estate) and reducing the barriers to entry of disruptive forces, such as start-ups.

- **Continual communication of city strategy performance** – Publicly demonstrating performance against governmental strategy in an easily digestible and consumable manner to drive greater trust and transparency between the government and its citizens.
- **Knowledge transfer and city legacy** – Putting in place initiatives in which solutions aimed at improving quality of life are exported pro-bono to developing nations as a means of reinforcing to next-gen talent the city's purpose, reach and influence.
- **Narrowing the digital divide** – The acceleration of Dubai's digital inclusion policy realization associated with the digital divide and ensuring greater parity in access to digital resources, tools and training amongst all resident groups.
- **Cascading technology top-down** – Cascading proven emerging tech use cases (i.e., blockchain) from the government level and through the value chain.
- **Decarbonization** – Becoming a global front-runner in the race to carbon net-zero.
- **Data leverage** – Ensuring the integrity and accessibility of open data platforms in order for the greater proliferation of data analytics solutions that provide both state, institutional and personal investors a better level of transparency and, therefore, encourage greater investment into the city.
- **The transition to a city-as-a-platform** – How will the city of Dubai continue to enact and delegate agile governance, offer integrated government services, encourage greater citizen participation and provide open access to data all through the means of an integrated, tech-enabled platform?

Sector-level initiatives

From the examples cited at a sectoral level, it is clear that, similar to the governmental level, there are a number of inspirational, purpose-driven initiatives in place designed to evolve the industries that will help drive the evolution of Dubai over the next 50 years. It is clear that the sectors that have helped fuel the diversification of the Dubai economy to date, such as energy and mobility (and more recently finance through Dubai International Finance Centre (DIFC)) are placing innovation at the heart of new business models that will fuel growth in the future.

Looking through the lens of the enabling framework outlined in Section 3 of this chapter, and building on the sectoral case studies, there are a number of key challenges and opportunities proposed that if addressed, could further enhance the effectiveness of sector-level capability and transformation – the outcomes of which could prove to be even more successful in helping Dubai achieve its vision.

- **Embedding the principles of Dubai's vision throughout sector value chains** – How effectively can large organizations and quasi-governmental entities positioned at the top of sectoral value chains embed, incentivize and contractualize the key principles and pillars of Dubai's city vision (i.e., city happiness, sustainability, digital transformation, human-centric innovation, collaboration, tolerance, the leverage of new technology) amongst their value chains?

- **Measurement of industry change effectiveness** – Building on their sustainability, innovation strategies and transformation agenda, what are the tangible performance measures in place to measure their effectiveness, and how can these measures be cascaded throughout value chains? How is this positioned within an overall Environmental, Social and Governance ("ESG") framework?
- **Evolution of procurement models and contracting mechanisms** – Moving beyond traditional, adversarial and transactional relationships with supply chains and instead adopting new collaborative and relational procurement and contracting approaches that support the co-creation of new value and hence drive innovation and industry transformation.
- **Exploration of vertical integration to drive and accelerate transformation** – The investigation and stimulus of vertical integration measures within sectors to support better collaboration, drive efficiency, more effectively incubate innovation and accelerate digital transformation.
- **Cross-sector collaboration** – How can new thinking, capabilities and solutions be leveraged across different sectors to accelerate change and better scale their benefits?
- **Transition to platforms that reduce barriers to entry** – How can sectors and industry create and act as integrated platforms to support business establishment, and hence reduce barriers to entry and transaction cost, give better access to new forms of finance (an objective of DIFC) and incentivize the growth of new ecosystems that bring innovation and value growth?
- **Enhanced operational agility** – How do organizations and sectors evolve their operations and transition to "scaled agile" models to more effectively respond to changes in market dynamics, better leverage new supply chain capabilities and their ability to harness industry disruptions?
- **Attracting, nurturing and retaining talent** – At a time when a shortage of talent has never been so profound, how do organizations improve their employee value propositions? What strategies are being implemented in terms of health and well-being, mentoring, future-focused skills acquisition, hybrid working models and workplace transformation?
- **Building the business case for decarbonization** – How effectively can sector stakeholders embed decarbonization as a key pillar of their business strategies and approach to ESG?

Education and cultural-level initiatives

Like the initiatives and strategies cited at National, Government, City and Sectoral, Dubai is clearly creating solid foundations to support the sustainable evolution of skills and knowledge transfer through collaborative approaches, whilst putting in place thoughtful and sensitive policy aimed at embracing multi-culturalism that will, in turn, drive the principles of inclusivity and human-centrality.

With the above in mind, some of the challenges and opportunities specific to education and culture could be:

- **Holistic approaches to skills development** – How can the various levels of required skillset be nurtured in an inclusive and industry-integrated manner? More specifically, and building on initiatives such as DFF's 100,000 coders, how can critical industry-required vocational skills be evolved and how will support be provided for those wishing to change careers due to automation, AI and industry other disruptions in order to retain talent in Dubai?
- **Industry-focused knowledge sharing** – As an outcome of public, private and educational partnerships such as those identified above at Heriot-Watt University, how will knowledge and solutions be made accessible across industries and be monetized in order to accelerate sector transformation? How can educational institutions act as facilitators of cross-industry knowledge sharing?
- **International and local educational institution collaborations** – How will established international educational institutions collaborate with their local Dubai-based counterparts in order to support the development of a world-leading educational infrastructure that organically produces next-generation talent?
- **Industry support to scale innovations conceived by educational institutions** – How can key industrial and public sector stakeholders better support the identification of use cases, funding and scaling of innovative solutions that will drive positive change, i.e., the use of graphene (a next-generation industrial material identified by at the University of Manchester), 3D printing, additive manufacturing.
- **Embedding tolerance in organizational ESG frameworks** – Building on the UAE's National Tolerance Program, how will its principles be embedded within corporations and their evolving ESG frameworks to promote cultural inclusiveness, human-centricity and enhanced community engagement?

Community-level initiatives

- **Scaling and monetizing integrated and sustainable community concepts** – How can the longer-term investment models of integrated sustainable communities such as The Sustainable City identified above, act as the model for community development in the future? How can the property investment and real estate market be evolved in a manner that goes beyond the build-it-and-they-will-come model and places a greater emphasis on medium- to longer-term residential investors and hence retains talent in Dubai?

- **Data-driven investment cases** – Given the availability of data in a Dubai's real estate market now established for over 20 years and acknowledging the proliferation of accessible advanced analytics technology, machine learning and AI, how can data-driven investment cases be developed that permit greater access to development finance and funding and preferential funding mechanisms for home purchasers?
- **Incentivization of multi-tenanted, integrated and sustainable communities** – In addition to demonstrating the favourable financial conditions for locating in new integrated community concepts such as The Sustainable City and Expo City, how can we further incentivize corporate and local small- and medium-sized enterprise (SME) businesses, start-ups, education providers and residents to locate and buy-in to desirable districts and communities and hence create occupational and institutional demand to invest in Dubai?
- **Citizen participation in masterplan 2040 delivery** – How can agile approaches to community development be employed to harness citizen (end-user) participation more effectively in the design and delivery of new differentiated community-focused real estate products and hence create more resiliency and drive greater sustainability in the real estate market?
- **Government support for industry transformation** – Given the opportunities above, how can the industry transformation required of the construction and real estate sector (a sector struggling due to high levels of fragmentation and low margins) be better supported and accelerated by the Dubai government to enable the delivery of modern, resilient and integrated communities aligned to the Dubai 2040 masterplan?

Opportunities for differentiation

It is clear from the previous section that Dubai is making tremendous strides in its quest to become the world's happiest city through a focus on the needs of its residents and visitors. It is also evident that the necessary strategies and initiatives that align and are supportive of this vision are being implemented across multiple sectors.

So far in this chapter, a number of key generic enablers responsible for driving Smart City growth (Section 3) have been identified, with a selection of Dubai's evolutionary strategies and initiatives positioned and reviewed through the lens of these enablers (Section 4). As an outcome, a number of specific challenges and opportunities were proposed, and as such provide the basis through which the key opportunities for Dubai to genuinely differentiate itself on the world stage can now be positioned.

These opportunities have been grouped and are summarized according to the five main areas of focus as follows.

1   **Decarbonization**

No one will argue, that Dubai's growth has largely been fuelled and has benefited significantly from revenues raised through the export of hydrocarbons. Whilst Dubai has been hugely successful in gradually weaning itself off a reliance of this source of income (it now demonstrating significant levels of differentiation in its economy), it has significant potential in becoming a champion of, and world leader in developing innovations that support decarbonization and the world's race to Net Zero – in a way, outlining the city's ability to showcase circularity and using its visibility for the good of the planet and position itself at the centre of global energy transition.

In addition to building on sector-led initiatives such as Warsan 2 and MBR Solar Park, it is important to acknowledge that Dubai is identified on the United Nations Framework Convention on Climate Change (UNFCCC)'s NAZCA web portal as a committed stakeholder enacting legitimate climate action, this is especially encouraging given that portal is in place to drive accountability for members and to demonstrate progress. It has also been announced that Dubai will host the UNFCCC's 28th session of the Conference of the Parties (COP 28) at Dubai Expo City in November 2023 – a key event that provides an opportunity for Dubai and UAE to showcase its progress and support against the Paris Agreement, promote international action and draw upon the legacy and vision of Expo 2020, "Connecting Minds, Creating the Future".

At a national level, it has been reported that the UAE is also emerging as one of the world's biggest state financiers of clean energy and has, "committed $400 million to enable developing nations' transition to clean energy and pledged to help supply green electricity to 100 million Africans by 2035" (Emirates News Agency, 2022).

A laser focus on decarbonization, however, will not only demonstrate and reinforce Dubai's position in a global community that is active in the fight against climate change, positioning decarbonization as a key pillar within the wider sustainable agenda of an "ESG" framework for governmental, public, private and institutional stakeholders will drive urgency and focus around organizational transformation with targeted outcomes including; zero waste, societal inclusiveness, workplace well-being, sustainable food systems, and zero carbon buildings, which will in-turn provide the robust use-cases for wider innovation and the leverage of technology for increased resilience and agility.

The effectiveness of a potential focus on decarbonization that drives a more general and balanced approach to sustainability does however rely upon alignment and integration of strategies amongst an array of different stakeholders.

2   **Strategy harmonization**

As outlined in the previous section, the government of Dubai has assembled and conceived an impressive portfolio of strategies and initiatives that will support and shape the city over the next 50 years. Rather than operate independently, Dubai has a real opportunity to integrate and leverage the synergies that enhanced collaboration and cooperation could bring.

More specifically and given that the start-up ecosystem is increasingly emerging as a driving force of the region's economic and social development, and building on the case studies presented in Section 4, it is apparent that the co-existence of structured approaches to externally sourcing innovation and open ideas platforms that encourage wider citizen participation are gaining more traction at government, educational and corporate levels. Through consolidation and transition to an integrated, tiered, connected city-wide platform for citizen engagement and innovation, there is an opportunity to ensure ideas are not missed, ideas can be joined up for scale and more effectively sign-posted to incubation and mentoring by the appropriate stakeholders and established resources and capabilities (such as those developed by Dubai Future Foundation, for example).

Likewise, and illustratively a citizen with a suggestion or comment about a DEWA government service could be signposted to potential collaboration opportunities with an innovation that was already being incubated with Heriot Watt University's Centre of Excellence for Smart Construction or Dubai Future Foundations' Hub 71 initiative.

Through demonstrating that ideas will not simply get dismissed, nor lost, citizen engagement could be more effective and motivate individuals from all backgrounds to get involved and have a stake in their city and support its vision to be the happiest and most human-centric place in the world. Powered by increasingly accessible AI technologies (accommodated and scaled more effectively through an agile approach to regulation and governance spearheaded by Digital Dubai) such integrated platforms could be extremely effective in accelerating innovation, whilst being a genuine tool for city-wide inclusivity.

Taking inspiration from, and similar to Dubai's participation in the UNFCCC's NAZCA Global Action Portal, Dubai also has a significant opportunity to be open around the performance of their happiness and Smart City vision through publicly accessible data and key performance indicators, which can be linked to the efforts of various stakeholders across the city. Such performance trends and analysis could be used to incentivize start-ups, citizens, SMEs and corporates alike to come forward with suggestions and ideas to accelerate efforts and outcomes in key areas. This could also be linked to the integrated engagement and innovation platform highlighted above and at a broader level be used as a mechanism to drive accountability, transparency and trust across the city (The Sustainabilist, 2019).

3   Global networks and the city-as-a-platform

Dubai has the potential to become an influential actor, or node, in an increasingly connected world that will be defined by emerging cross-border collaborations and partnerships. This will be especially important as Dubai looks to reinforce its position as a hub for global trading and logistics and leverage its position as facilitator and orchestrator of trading flows between East and West – this vision being reinforced by Dubai's support for the Belt and Road Initiative (BRI) which was announced by the Chinese Government in 2013 with the aim of developing the emerging Asian economies and strengthening their trade and economic relations with the rest of the world, coinciding with Dubai's own efforts at reviving the Silk Road trading route as a key trading gateway into China.

New forms of collaboration and sectoral ecosystem partnerships will require transactions between partner stakeholders to be automated and seamless, and to be formed in an agile manner. There is, therefore, a significant opportunity to enable this more effectively if it continues to evolve its city-as-a-platform approach in order to support business establishment, hence reduce barriers to entry and transaction cost, give better access to new forms of finance (an objective of DIFC) and incentivize the growth of new ecosystems that bring innovation and value growth.

Not only would this standardized user experience allow citizens, businesses, start-ups and educational institutions to "plug themselves" into Dubai but also allow other cities and maybe other entities comprising the MENAP region to collaborate and exchange knowledge, open up previously disintegrated markets and drive the operationalization of city-specific initiatives such as Digital Dubai's Smart Cities Global Network (as outlined in Section 4).

4   Sector, skills and workplace transformation

The role of private enterprise working in alignment and in collaboration with city government has been proposed as a key enabler in the evolution of a Smart City. The key opportunities for sectoral transformation and to the organizations that sit within them are; to wholly align with the Dubai vision, to ensure corporate strategies contain objectives and initiatives to accommodate and drive change, to harness knowledge and skills outside of their own sectors and industries and mandate and measure the impact of their transformation agenda throughout their value chain.

To leverage new technologically enabled capabilities and the increased value that comes with them, organizations also need to become more agile in their operational models in order to more effectively incubate, enable experimentation, procure and scale new propositions that have been conceived thanks to the city's relentless focus on innovation. As already suggested, there are significant opportunities for organizations to work collaboratively, and cross-sector in order to reach their objectives more quickly

More specifically, some, more traditional low-margin industries involving high levels of fragmentation in their supply chains, such as

Construction and Real Estate ("CRE"), have an acute opportunity to evolve their procurement models and contracting mechanisms in a manner that moves beyond traditional, adversarial and transactional relationships with their suppliers and instead adopt new collaborative and relational procurement and contracting approaches that supports the co-creation of new value, a more equitable allocation of risk, incentivizes integrated supply chain solutions and hence drive innovation and industry transformation. The support of the Dubai government may be required in order to mandate this requirement.

Given that the CRE sector will be responsible for delivering the physical spaces where Dubai city dwellers work, live and play and will therefore be a critical stakeholder in the evolution of the city and its achievement of the 2040 masterplan; a specific focus should be placed on its transformation especially given that despite advances in technology, productivity has barely increased over the last 30 years and not kept pace with overall economic productively (Agarwal et al., 2016). More specifically opportunities can be viewed through the lens of decarbonization, talent and worker well-being.

- **Decarbonization** – Given that construction is the largest global consumer of raw materials, and the built environment accounts for 25%–40% of the world's total carbon emissions (The World Economic Forum; The Boston Consulting Group, 2017), there is a significant opportunity and responsibility of the industry to improve. Taking inspiration from the UNFCCC and industry-backed educational institution R&D initiatives (such as the Heriot-Watt Centre of Excellence in Smart Construction), the city of Dubai working in collaboration with the industry has the following specific opportunities that, in turn, will drive much-needed transformation:
    - Private sector industry and governments to co-create strategies and delivery plans aimed at considerably reducing embodied emissions in infrastructure projects, new build and refurbishment projects, and achieve net-zero emission construction sites.
    - Develop and mandate procurement policy that places an emphasis on the specification of low carbon materials and zero emission construction plants.
    - Encourage (through incentivization), the use of low carbon materials, the establishment of net-zero construction sites and utilization of resource efficient and circular design.
    - Embedding Life Cycle Assessments (LCA's) in planning policy.
    - Impact assessments of material and design choices on the overall resilience of the city (and its ability to respond) to climate change.
- **Talent** – The industry faces a significant skills shortage driven by an ageing workforce. Technology implementation can be one way to combat this (especially given that construction is one of the least digitized), but more acutely, the industry needs to develop an impactful

purpose to attract next-generation talent and the individuals that will drive its transformation. Collectively, the UAE Centennial Plan 2071, Dubai 2040 Masterplan, Dubai Future Foundation, Digital Dubai (and others) and the call to action specific to decarbonization above, provide the perfect platform to genuinely inspire new entrants to the industry.

- **Well-being** – Gen X (millennials), Gen Z and future-generation talent are also placing more emphasis on non-financial remuneration such as hybrid/flexible working models, diversity and inclusion, health and well-being – all elements that CRE in Dubai can revolutionize as part of a wider government-backed and collaborative transformation.

5 The emergence of new physical spaces – What will Dubai look like in 50 years?

Given the evolutionary and inspired Dubai city strategies and initiatives that have been put in place, the successful navigation of key challenges and the leverage of the significant opportunities Dubai has as outlined above, what could the physical form of Dubai begin to look like over the next 50 years?

On the basis of the evolution of empowered industries with new capabilities to deliver, an interactive and human-centric masterplan that enables citizen participation, agile planning policy, data-driven and sustainable business cases for real estate investment and new forms of financing, some of the key characteristics of Dubai could include:

- **The rise of the 20-minute neighbourhood** – Smart, integrated communities that are within a 20-minute walk, cycle or public mode of transport of all necessary facilities used for everyday life.
- **Integrated platform communities** – The rise of platforms that serve the needs of citizens, business tenants and residents that reside for the benefit of the community. This may include seamless data sharing around energy consumption and production, automated community management service provision, new community start-up services, home purchase support and financing, all designed to drive community engagement and connection with the rest of Dubai.
- **Net zero/Off-grid communities** – In line with communities such as The Sustainable City, Dubai, solar energy production within all communities to enable them to be entirely self-sufficient.
- **Tenant co-existence** – An increase in integrated communities in which start-ups, local businesses, corporates, residents, educational institutions co-exist in response to innovative working models and practices.
- **Community focus** – An increase in multi-cultural and multi-demographic groups living in mixed neighbourhoods in line with the principles of the UAE's National Tolerance Program and social inclusion, together with the provision of community-level

education and research centres aimed at empowering individuals and supporting the development of innovative solutions to global challenges.

– **Increase in owner-occupation** – Given new forms of data-led financing and a further evolution of the real estate market, home ownership will likely increase as more and more people seek to settle in Dubai in the medium to long term.

– **Modular and offsite manufacture** – As the CRE industry becomes less fragmented and more industrialized/automated, there will be a greater opportunity for asset owners to become part of the configurability and customization process as part of a new agile, customer-centric approach to asset delivery.

– **Experimental spaces** – New technologically enabled experimental pop-up retail and commercial products designed as proof-of-concept in which feedback is sought and business models validated

– **Expansion and integration of transport network** – A further evolution of Dubai's integrated multi-model transport network that introduces new rapid forms of mobility such as Hyperloop, autonomous taxi drones and sharing of an electrified vehicle fleet.

– **Green space and urban farms** – A proliferation of green space and community-managed technologically enabled urban farming space.

– **Smart/Flexible buildings** – Buildings that are self-sufficient, net-zero, responsive to the changing needs of the market and the environment. Key features may include seamless and continuous, data-led re-configuration, self-repair through next-generation building materials, optimal energy production and consumption through advanced building management systems and digital twin technology.

– **Community-level education and research centres** – Educate people around taking responsibility for their environment and sustainability.

Conclusions

To get to where Dubai has today demonstrates vision and ambition that was propelled through fully leveraged market forces that were shaping world cities at the time. Human capital and global investment and finance were drawn to Dubai almost unhindered to deliver against its bold and unrelenting mission. We are entering the most decisive era that humankind has probably ever faced if we care to continue to live our lives in a manner to which we have become accustomed. Climate change, rapid urbanization, global food security and the war for talent are no longer risks, they are active issues and forces that are now pushing cities to respond and adapt in a manner that strengthens their agility and resiliency. To do this effectively will mean the leverage of new resources and be led by a newfound, collective and critical purpose that is founded on the principles of mutual cooperation, collaboration and knowledge sharing that as such, results in city outcomes that are able to benefit broader levels of society beyond its boundaries.

This chapter began with a proposition that whilst data and technology are an important component of Smart City growth, they are purely an enabler to driving outcomes focused on improving quality of life. Through the lens of the suggested Smart City enablers, the case studies presented and the opportunities that the city has to evolve in a differentiated manner, it is clear that Dubai is looking to its next 50 years in a balanced, inclusive, human-centric, innovative and connected manner. No one can dispute that Dubai is on an exciting trajectory and has every chance of delivering against its aims of becoming the happiest and smartest city on the planet provided it tackles the challenges and leverages the opportunities ahead.

Indeed, Dubai has every opportunity to become an influential actor, or node, in an increasingly connected world where it is able to fully leverage its position as a global trading hub where East meets West. Technology will allow Dubai to seamlessly engage, transact, collaborate and share knowledge and capabilities through a city platform approach that acknowledges the convergence of the physical and virtual worlds. The Dubai customer experience will transcend geographic boundaries and bring the city to the world in an immersive and inspiring manner.

But what about talent? Isn't that the true fuel that will power the engine of city growth? In its rapid growth stage, Dubai was defined by its transient, mobile and expat population in which talent exited as quickly as it arrived. As outlined in previous chapters of this book, through the articulation of a compelling city vision and the support of an evolving and maturing real estate sector, it is evident that Dubai is now offering a longer-term resident proposition that offers greater financial security and a feeling of belonging through integrated communities and active citizen participation programs – things that get people to stay.

This chapter has deliberately resisted comparing Dubai to other so-called Smart Cities (there are numerous city case studies and indices that are available, all with a different view). Dubai is in a unique position in which it has every opportunity to shape its destiny. Indeed, in a volatile and ever-changing world, comparison and benchmarking would be akin to looking back in time for the answers.

The key argument proposed in this final chapter is that whilst data and technology are an important component of Smart City growth, they are purely an enabler to driving outcomes focused on improving quality of life. Indeed, one must take a holistic, balanced (and increasingly human-centric) view if we are to truly gain an informed insight as to the key drivers that will continue to boost Dubai's city growth and urban development success story in the future. Dubai and its real estate market have plenty more to look forward to.

References

Agarwal, R., Chandrasekaran, S., and Sridhar, M. (2016) Imagining Construction's Digital Future, McKinsey and Company. https://www.mckinsey.com/business-functions/operations/our-insights/imagining-constructions-digital-future.

Aguiar, M., Boutenko, V., Lacanna, S., Mlodik, E., and Williams, M. (2020) Vibrant Cities Are Built on Trust, BCG Publication. https://www.bcg.com/publications/2021/vibrant-cities-are-built-on-trust#designing-for-trust.

BCG. Cities of the Future. https://www.bcg.com/industries/public-sector/cities-of-the-future.

Emirates News Agency (2021) Mohammed Bin Rashid Launches Dubai 2040 Urban Master Plan. https://www.wam.ae/en/details/1395302917640.

Emirates News Agency (2022) 'The Hottest Investor in Renewables Is a Big Oil Producer', edited by Mohamed, H. and Nashar, K. WAM (Emirates News Agency). https://www.wam.ae/en/details/1395303061717

Grandview Research (2022) Smart Cities Market Size, Share & Trends Analysis Report By Application, By Smart Governance, By Smart Utilities, By Smart Transportation, By Region, And Segment Forecasts, 2022–2030 (Report ID: 978-1-68038-270-9). https://www.grandviewresearch.com/industry-analysis/smart-cities-market

McKinsey Global Institute (2018) 'Smart Cities Digital Solutions for a more liveable future' https://www.mckinsey.com/~/media/mckinsey/business%20functions/operations/our%20insights/smart%20cities%20digital%20solutions%20for%20a%20more%20livable%20future/mgi-smart-cities-full-report.pdf

MAF (2022) 'The time is now: A perspective on economic integration in MENAP', *Majid Al Futtaim and the World Economic Forum*, May 2022

Patrão, C., Moura, P., and Almeida, A.T. (2020) 'Review of smart city assessment tools', *Smart Cities*, 3, pp. 1117–1132.

Senge, P.M. (2006) *The Fifth Discipline: The Art and Practice of the Learning Organization*, New York: Doubleday

The Sustainabilist (2019) The Silk Road to Dubai's Green Gateway. https://thesustainabilist.ae/the-silk-road-dubais-green-gateway-to-the-world/.

Trindade, E.P., Hinnig, M.P.F., and da Costa, E.M. (2017) 'Sustainable development of smart cities: A systematic review of the literature', *Journal of Open Innovation: Technology, Market and Complexity*, 3 (11), p. 11.

UNEP (2018) *The Weight of Cities–Resource Requirements of Future Urbanization*, Paris: International Resource Panel Secretariat.

United Nations (2022) Department of Economic and Social Affairs: Sustainable Development, "The 17 Goals" https://sdgs.un.org/goals

Weber, V. (2019) Smart Cities Must Pay More Attention to the People Who Live in Them, World Economic Forum. https://www.weforum.org/agenda/2019/04/why-smart-cities-should-listen-to-residents/.

World Economic Forum (2017) Shaping the Future of Construction Inspiring Innovators Redefine the Industry, World Economic Forum and Boston Consulting Group. https://www3.weforum.org/docs/WEF_Shaping_the_Future_of_Construction_Inspiring_Innovators_redefine_the_industry_2017.pdf.

Zakzak, L. (2019) Citizen-centric Smart City Development: The Case of Smart Dubai's "Happiness Agenda", Proceedings of the 20th Annual International Conference on Digital Government Research, June 2019, pp. 141–147.

Zhu, H., Shen, L., and Ren, Y. (2022) 'How can smart city shape a happier life? The mechanism for developing a Happiness Driven Smart City', *Sustainable Cities and Society*, 80. https://doi.org/10.1016/j.scs.2022.103791.

References to external sources relating to key initiatives discussed in Ch.12

National, government and city level initiatives

- Digital Dubai, www.digitaldubai.ae.
- Digital Inclusion and Participation, https://u.ae/en/about-the-uae/digital-uae/digital-inclusion-and-participation
- Dubai 2040 Masterplan, UAE Government, http://www.dubai2040.ae/en/urban-master-plan.
- Government with You, https://u.ae/en/participate.
- Invest in Dubai, https://invest.dubai.ae/.
- MBR Majlis, www.mbrmajlis.ae.
- The Dubai Future Foundation ("DFF"), www.dubaifuture.ae.
- UAE Centennial 2071, UAE Government, www.u.ae.

Initiatives at the sector level

- AviationX Lab, https://www.theaviationxlab.com.
- Dubai International Finance Centre ("DIFC") – Open Finance Lab.
- Mohammed bin Rashid Al Maktoum Solar Park, www.mbrsic.ae.
- Roads and Transport Authority – Innovation Agenda, https://www.rta.ae/wps/portal/rta/ae/home/about-rta/innovation.
- Warsan 2 Waste to Energy Plant, https://www.besix.com/en/projects/dubai-waste-to-energy.

Initiatives at the educational and cultural levels

- Heriot Watt University in Dubai – Centre of Excellence in Smart Construction, https://www.hw.ac.uk/dubai/research/smart-construction.htm.
- Roads and Transport Authority – Research and Innovation Centre at the University of Birmingham Dubai, https://www.birmingham.ac.uk/news/2022/rta-partnership-creates-innovation-hub.
- The National Tolerance Program, https://u.ae/en/about-the-uae/strategies-initiatives-and-awards/federal-governments-strategies-and-plans/national-tolerance-programme.

Initiatives at the community level

- Expo City Dubai, https://www.expocitydubai.com.
- The Sustainable City Dubai, https://www.thesustainablecity.ae/.

Other weblinks

BESIX, Dubai Waste to Energy, https://www.besix.com/en/projects/dubai-waste-to-energy.

Cities Race to Zero, https://www.c40knowledgehub.org/s/race-to-zero-pledge-form?language=en_US.

Digital Dubai Authority, The Vision of H.H. Sheikh Mohammed Bin Rashid Al Maktoum Is to make Dubai the World's Digital Capital, https://www.digitaldubai.ae/about-us.

Dubai Future Foundation – Shaping the Future of Dubai, https://www.dubaifuture.ae/.

Government of Dubai, RTA's E3 Clinches GovTech Innovation Award, https://www.mediaoffice.ae/en/news/2022/June/13-06/RTAs-EC3-clinches-GovTech-Innovation-Award.

IBM (2022), What Is Industry 4.0 and How Does It Work? https://www.ibm.com/topics/industry-4-0.

ICMA, The City as a Platform, https://icma.org/articles/article/city-platform.

IMD (2021), Smart City Observatory: What Makes a City Liveable and Smart? https://www.imd.org/smart-city-observatory/Home/.

Oracle, What Is a Smart City? https://www.oracle.com/industries/government/smart-cities/what-is-a-smart-city/.

RTA, Innovation in the RTA, https://www.rta.ae/wps/portal/rta/ae/home/about-rta/innovation.

UAE Cabinet, The National Strategy for Innovation, https://uaecabinet.ae/en/the-national-strategy-for-innovation.

UAE Government Portal, Dubai 2040 Urban Master Plan, https://u.ae/en/about-the-uae/strategies-initiatives-and-awards/local-governments-strategies-and-plans/dubai-2040-urban-master-plan.

UAE Government Portal, Dubai Data Strategy, https://u.ae/en/about-the-uae/strategies-initiatives-and-awards/local-governments-strategies-and-plans/dubai-data-strategy.

UAE Government Portal, UAE Centennial 2071, https://u.ae/en/about-the-uae/strategies-initiatives-and-awards/federal-governments-strategies-and-plans/uae-centennial-2071.

UAE Government Portal, UAE Energy Strategy 2050, https://u.ae/en/about-the-uae/strategies-initiatives-and-awards/federal-governments-strategies-and-plans/uae-energy-strategy-2050.

UAE Ministry of Finance, 2020: Towards the Next 50, https://mofsystems.mof.gov.ae/innovativereleases/towards-the-next-50.html.

Index

Note: **Bold** page numbers refer to tables and *italic* page numbers refer to figures.

Printed in the United States
by Baker & Taylor Publisher Services